Springer Theses

Recognizing Outstanding Ph.D. Research

Aims and Scope

The series "Springer Theses" brings together a selection of the very best Ph.D. theses from around the world and across the physical sciences. Nominated and endorsed by two recognized specialists, each published volume has been selected for its scientific excellence and the high impact of its contents for the pertinent field of research. For greater accessibility to non-specialists, the published versions include an extended introduction, as well as a foreword by the student's supervisor explaining the special relevance of the work for the field. As a whole, the series will provide a valuable resource both for newcomers to the research fields described, and for other scientists seeking detailed background information on special questions. Finally, it provides an accredited documentation of the valuable contributions made by today's younger generation of scientists.

Theses are accepted into the series by invited nomination only and must fulfill all of the following criteria

- They must be written in good English.
- The topic should fall within the confines of Chemistry, Physics, Earth Sciences, Engineering and related interdisciplinary fields such as Materials, Nanoscience, Chemical Engineering, Complex Systems and Biophysics.
- The work reported in the thesis must represent a significant scientific advance.
- If the thesis includes previously published material, permission to reproduce this must be gained from the respective copyright holder.
- They must have been examined and passed during the 12 months prior to nomination.
- Each thesis should include a foreword by the supervisor outlining the significance of its content.
- The theses should have a clearly defined structure including an introduction accessible to scientists not expert in that particular field.

More information about this series at http://www.springer.com/series/8790

Matej Blatnik

Groundwater Distribution in the Recharge Area of Ljubljanica Springs

Doctoral Thesis accepted by
Graduate School, University of Nova Gorica,
Nova Gorica, Slovenia

 Springer

Author
Dr. Matej Blatnik
Karst Research Institute
Research Centre of the Slovenian
Academy of Sciences and Arts
University of Nova Gorica
Postojna, Slovenia

Supervisors
Prof. Franci Gabrovšek
Karst Research Institute, Research Centre
of the Slovenian Academy
of Sciences and Arts
University of Nova Gorica
Postojna, Slovenia

Prof. Mihael Brenčič
University of Ljubljana
Ljubljana, Slovenia

ISSN 2190-5053   ISSN 2190-5061 (electronic)
Springer Theses
ISBN 978-3-030-48338-8   ISBN 978-3-030-48336-4 (eBook)
https://doi.org/10.1007/978-3-030-48336-4

This Springer imprint is published by the registered company Springer Nature Switzerland AG
The registered company address is: Gewerbestrasse 11, 6330 Cham, Switzerland

Supervisor's Foreword

Characterisation of spatial and temporal distribution of groundwater in karst aquifers remains one of the main challenges in karst hydrogeology. Namely the flow in mature karst aquifers is focused on networks of solution conduits, whose distribution and geometry are largely unknown. Different approaches have been used to address the problem, from direct surveying of accessible conduits to geophysical and hydrological techniques. Flow patterns are even harder to predict in young orogens, where complex structures and active tectonics dictate extremely segmented and unpredictable distribution of conduits. The classical example of such aquifers is in the Dinaric karst.

This thesis presents a promising approach towards understanding the relation between groundwater dynamics and the structure of karst aquifers. It is based on data sets obtained from long-term continuous monitoring of basic groundwater parameters in epiphreatic caves, ponors, and springs in the recharge area of the Ljubljanica Springs, Slovenia. The research area is one of the prominent classical karst sites, where karst surface, caves, and groundwater flow have been studied and explored for centuries. The obtained data were interpreted in view of complementary speleological, geological and hydrological data, and basic hydraulic principles. Heuristic conceptual models of local and regional aquifer structure were converted to hydraulic models. By iterative comparison of modelled and observed responses to flood events, the models were validated and updated.

The results of this work give insights into the flood response of the observed system, predict the yet unknown structure and flow pathways, and elucidate the interaction between the karst polje and adjacent aquifer.

The implications of this work for karst hydrogeology are manifold. It gives practical insights into how and where to set up monitoring sites in caves. It develops a methodology, which integrates speleological and hydrological data and demonstrates the use of a simple hydraulic model in data interpretation.

The complexity of this research demanded many skills and a broad knowledge. It required regular maintenance and data transfer at observation sites in eight caves. These are up to 300 m deep and reaching them takes several hours of climbing and crawling. To better constrain the locations and local geometry, new cave surveys have been done. As a keen caver, Matej Blatnik already had the field skills, which was in this case almost a prerequisite. In order to complete the research, he acquired a profound knowledge of analytical methods, speleology, hydraulic modelling, and regional hydrogeology. As a supervisor, I admired the progress that Matej made in all these topics during his doctoral study and research. I enjoyed discussions with him and many common field trips to the caves. Last but not least, luckily, nature has provided enough characteristic flood events to analyse.

Vrhnika, Slovenia   Prof. Dr. Franci Gabrovšek
April 2020

Abstract

The peculiar hydrology of the karstic recharge area of the Ljubljanica River has attracted researchers for centuries. At the beginning of the nineteenth century, the research focused on preventing the flooding of Planinsko Polje. A large research project in the 1970s resulted in the delineation of main water pathways in the recharge area. However, the dynamics of the groundwater and its relation to aquifer geometry has not been studied until recently. New speleological discoveries in addition to the recent development of automatic instruments with the internal memory now allow for the distributed observation of groundwater with a high temporal resolution.

The purpose of the present study is, therefore, to improve the knowledge of the groundwater dynamics in the recharge area of the springs of the Ljubljanica River through the use of the recently improved techniques for observation and interpretation.

The study is focused on the northern part of the Ljubljanica River recharge area, between Planinsko Polje in the south and the springs of the Ljubljanica River in the north. There, a network with autonomous programmable instruments was established. Observations of water level, temperature, and specific electrical conductivity took place at four ponors on Planinsko Polje, three springs of the Ljubljanica River, and eight water-active caves between the ponors and the springs. Additionally, numerous field measurements of water parameters were collected, and a large amount of hydrological and meteorological data from the Slovenian Environment Agency were obtained and used for the analysis.

The newly obtained data were compared and analysed based on previous knowledge of the system. Many previous findings were confirmed, but also new interpretations were developed. Interpretations related to the influence of underground geometry to water level variation were further tested using simplified conceptual and numerical models. For this, EPA Storm Water Management Model (SWMM) software was used.

During the 3.5 year study period, more than 15 flooding events on Planinsko Polje occurred. The high variability in the intensity and duration of the events resulted in a rich data set. The longest flood lasted for three months, and during this period, the water level in the entire aquifer was also at its highest; it rose up to 66 m.

The measurements confirm previous findings that show that water in the aquifer interweaves (diverges and converges), and that water from the northern ponors of Planinsko Polje mostly flows towards the western springs of the Ljubljanica River, whereas water from the eastern ponors on Planinsko Polje mostly flows towards the eastern springs of the Ljubljanica River. Water coming directly from Cerknisko Polje also has an important influence on the eastern springs, which has been recognised based on the acquired temperature signals.

Analyses of the water level hydrographs provide novel interpretations of the aquifer's geometry. At all observation points, the water level hydrographs show inflection points with a temporarily slower increase and decrease of the water level, which indicate the presence of overflow passages. Some of the overflow passages were already known (in caves Gradišnica, Logarček, and Najdena jama), whereas some were anticipated (in caves Veliko brezno v Grudnovi dolini and Andrejevo brezno 1).

The water level hydrographs also indicate a possible influence from geological structures. In cave Veliko brezno v Grudnovi dolini, an abrupt transition from rapid decrease to stable stage is related to damming by a hydrogeological barrier. A geological barrier positioned downstream can also result in simultaneous water level variation in consecutive caves (i.e. caves Gradišnica and Gašpinova jama). On the northern border of Planinsko Polje, redirection of groundwater flow due to a geological structure was also identified. A previously anticipated connection between ponor zone Pod stenami and cave Najdena jama was refused and explained by the bypassing of the expected flow and discharging through another group of ponors. Identification of possible vertically changeable transmissivity of the Idrija Fault Zone also assists in interpreting the function of estavelles near Grčarevec and back-flooding in caves located in the Hrušica Plateau (i.e. caves Veliko brezno v Grudnovi dolini and Andrejevo brezno 1). Different responses to the rain events indicate that the observation points are recharged from one or more water flow directions.

Water temperature and specific electrical conductivity measurements indicate periods when active water flow passes the observation points. During these periods, signals from the surface flow were transferred and time lags between signals in consecutive locations enabled calculations of transit times and further apparent water flow velocity. Sudden increases in specific electrical conductivity also indicate possible pollution of the water.

The large number of observation points, a long observation period, continuous autonomous measurements, and the use of new interpretation techniques including testing with numerical modelling provide a comprehensive analysis with many new findings. The results give a deeper view into the geometry of the aquifer, which is important because the aquifer has a strong impact on the water level dynamics.

Like the previous studies, this study also leaves many open questions remaining and indicates areas for further research. Future investigations may benefit by focusing on a more detailed study of the influence of the geological structure, a recognition of the phreatic and deep regional flow, an assessment of the water balance and water quality, and a generally better understanding of the flooding of Planinsko Polje.

Keywords   Karst aquifer · Groundwater dynamics · Flooding · Modelling · Planinsko Polje · Ljubljanica River

Parts of this thesis have been published in the following journal articles:

- Blatnik, M., Frantar, P., Kosec, D. & F. Gabrovšek, 2017: Measurements of the outflow along the eastern border of Planinsko Polje, Slovenia.- Acta Carsologica, 46, 1, 83–93. https://doi.org/10.3986/ac.v46i1.4774
- Ravbar, N., Petrič, M., Kogovšek, B., Blatnik, M. & C. Mayaud, 2018: High waters study of a classical karst polje - an example of the Planinsko Polje, SW Slovenia. In: Milanović, S. & Z. Stevanović (eds.): Proceedings of the International Sympsium KARST 2018 "Expect the Unexpected", 6–9. June 2018, Trebinje. Belgrade: Centre for Karst Hydrogeology; Trebinje: Hydro-Energy Power Plant "Dabar", 417–424.
- Mayaud, C., Gabrovšek, F., Blatnik, M., Kogovšek, B., Petrič, M. & N. Ravbar, 2019: Understanding flooding in poljes: a modelling perspective.- Journal of Hydrology, 575, 874–889. https://doi.org/10.1016/j.jhydrol.2019.04.092
- Blatnik, M., Mayaud, C. & F. Gabrovšek, 2019: Groundwater dynamics between Planinsko Polje and springs of the Ljubljanica River, Slovenia.- Acta Carsologica, 48, 2, 199–226. https://doi.org/10.3986/ac.v48i2.7263
- Blatnik, M., Culver, D.C., Gabrovšek, F., Knez, M., Kogovšek, B., Kogovšek, J., Liu, H., Mayaud, C., Mihevc, A., Mulec, J., Năpăruş-Aljančič, M., Otoničar, B., Petrič, M., Pipan, T., Prelovšek, M., Ravbar, N., Shaw, T., Slabe, T., Šebela, S. & N. Zupan Hajna et al. 2020: Deciphering epiphreatic conduit geometry from head and flow data. In: Knez, M., Otoničar, B., Petrič, M., Pipan, T. & T. Slabe (eds.): Karstology in the Classical Karst.- Springer series: Advances in Karst Sciences, 149–168. https://doi.org/10.1007/978-3-030-26827-5_8
- Olarinoye et al., 2020: World Karst Spring hydrograph (WoKaS) database for research and management of the world's fastest-flowing groundwater.- Nature Scientific Data, 7:59, 1–9. https://doi.org/10.1038/s41597-019-0346-5

Acknowledgements

This work could not have been completed without the help of numerous contributors.

First, I would like to thank both of my mentors, Franci Gabrovšek and Mihael Brenčič, for proposing an interesting topic that included an engaging combination of field work and accompanying analyses. I found their guidance, advice, and support through the research process to be incredibly helpful. Many thanks also go to the members of the commission, Metka Petrič, Steffen Birk, and Luca Zini, for their useful suggestions.

Thank you to all of the past and present collaborators at ZRC SAZU Karst Research Institute for our numerous discussions. Your suggestions, advice, and support were always welcome. And our work together, while helping each other with our various fieldwork projects, helped me to gain additional knowledge and experience.

Cooperation with researchers from other institutions was also very important during this research. Thanks to members of Slovenian Environment Agency, Slovenian National Building and Civil Engineering Institute, Geological Survey of Slovenia, Intrepid Geophysics, Swiss Institute for Speleology and Karst Studies, University of Nice, Tallinn University, University of Arkansas and Yunnan University.

This project required a lot of fieldwork. Although it was often relaxing to go out alone, in caves, where safety and physical assistance are necessary, each hand was greatly appreciated. For joining me on cave visits, sharing cave maps and other materials, and assistance with other cave-related activities, I would like to thank at least 50 cavers, members of following caving clubs: DZRJ Ljubljana, JD Rakek, JD Logatec, DZRJ Luka Čeč Postojna, JK Železničar, ASAK Beograd, CGEB Trieste, and SO PDS Velebit.

Thanks to Stephanie Sullivan for revising the English text.

I would also like to thank my parents, Olga and Alojzij, and my siblings, Damjana and Alojzij, for supporting me through my geography and karstology studies and for their enthusiasm for any work related to nature.

Last but not least, I would like to thank the Slovenian Research Agency (ARRS) for financially supporting my work as a young researcher.

Contents

Chapter 1
Introduction

1.1 The Extent and Importance of Karst: Global and Slovene Perspective

For the development of a karst landscape, two main things are necessary: soluble rock and water [11, 14, 15]. Slovenia is rich in both. Soluble carbonate rocks cover 47% of the surface [18]. This is significantly more than the global percentage, which ranges from 15 [16] to 20% [11, 19, 38] of Earth's ice-free continental area. Limestone constitutes approximately 60% of the carbonate rock in Slovenia, dolomite about 30%, and clastic carbonate rocks 10% [18]. The thickness of carbonate rocks in Dinaric Karst is up to several thousand metres. The area is in the plate collision zone and, therefore, tectonically very active, resulting in highly fractured rock and tectonically conditioned karst evolution [36]. The amount of precipitation ranges between 900 and over 3000 mm, with an average of 1450 mm [12], which is also more than the 1000 mm of average global rainfall [22]. A large amount of carbonate rocks and rain are, therefore, two main factors that enable well developed karst with dominant conduit drainage. Karst aquifers are among the most prolific water resources. Approximately 20–25% of the global population depends largely or entirely on groundwater [11], whereas, in Slovenia, the share is even higher at about 50% [4, 30]. The position of solution conduits, which carry most of the flow, is typically unknown. For this reason, although abundant, karst aquifers are highly vulnerable and often poorly characterised water resources. As such, they pose many challenges to the safety of the water supply.

© The Editor(s) (if applicable) and The Author(s), under exclusive license
to Springer Nature Switzerland AG 2020
M. Blatnik, *Groundwater Distribution in the Recharge Area
of Ljubljanica Springs*, Springer Theses, https://doi.org/10.1007/978-3-030-48336-4_1

1.2   Goal, Objectives, and Hypotheses of This Work

The **goal** of this research is to develop new methods and approaches for a better char-acterisation of karst aquifers based on autonomous measurements of groundwater parameters in ponors, caves, and springs.

The study area is located between ponors on Planinsko Polje and springs of the Ljubljanica River. Many studies of groundwater flow were made in this area [17, 32, 34]. The studies show a very dispersed and intertwined groundwater flow dependent on the structural and lithological diversity of the area. In spite of the studies, many questions were not solved.

There are many objectives, leading to reach the goal. Some of them are more general and are related to processes that are present in a karst aquifer, whereas, specific objectives are related to the selected study area.

Specific Objectives Related to the Study Area:

- to set up a monitoring network of autonomous measurements of groundwater parameters in ponors, caves, and springs;
- to interpret the obtained hydrographs of water level, temperature, and specific electrical conductivity in order to

  - relate hydrographs to the known geometry (i.e. size and position of epiphreatic conduits),
  - identify the flow paths in the observed system,
  - infer the unknown geometry between the subsequent observation points along the flow path,
  - relate groundwater dynamics to structural, lithological, and speleological elements,
  - study the relation between Planinsko Polje and the adjacent aquifer,
  - determine transit times between subsequent points;

- to construct simple conceptual and numerical hydrological models of the system.

General Objectives:

- to develop a comprehensive methodology for a better characterisation of karst aquifers based on the monitoring network, qualitative data interpretation, and numerical evaluation of conceptual models;
- to assess the importance and relevance of caves as observation points.

Hypothesis

A network of observation points, which includes caves with active groundwater flow, would provide a significantly deeper insight into the geometry and functioning of a karst aquifer, particularly in the epiphreatic zone, than limiting monitoring to the inputs and outputs of an aquifer.

1.3   Characteristics of a Karst Aquifer

An aquifer is a rock formation that stores, transmits, and yields an economically significant amount of water [10]. Karst aquifers owe their properties to the solubility of rocks in groundwater. In the case of carbonate karst, CO_2 dissolved in water drastically increases the solubility of calcite/dolomite in groundwater [1]. Groundwater dissolutionally widens initial discontinuities (bedding planes, fractures) in carbonate bedrock and thus forms a network of solution conduits [8, 28].

According to the nature of the voids in which the water is stored and transmitted, a *genetic* classification distinguishes three types of porosity in karst aquifers: primary, secondary, and conduit [11, 38].

- **Primary porosity** (also **granular or matrix porosity**) is characterised by pores at the smallest scale. These pores are distributed between grains or crystals in the rock and were formed during diagenesis by the packing of mineral grains in the rocks [10, 28, 38]. Granular porosity is the highest in young limestones and very low in old, well consolidated limestones, where pores were filled with cementing minerals. Other voids may be formed by solution during or after diagenesis [28]. Granular porosity varies from 2.4 to 30% [35, 40].
- **Secondary** (also **fractured) porosity** comprises joints, fractures, and bedding plane openings. They might have tectonic (fractures, joint) or sedimentary (bedding planes) origins. They form in all rocks, but massive brittle rocks such as sandstones, granites, limestones, and dolomites are better able to maintain the mechanical openings [38]. Their volume presents a porosity of up to about 3.5% [11].
- **Conduit porosity** is significant for karst aquifers, which are distinguished from all others by the presence of dissolution of carbonate rock and the development of large solution cavities and integrated conduits with a cross sectional area from several mm^2 to several 10 m^2 [5, 38]. Channel porosity varies from 0.003 to 0.5% [35, 40], but these openings often drain 99% or more of the groundwater flow [39].

In order to make a precise description, one has to consider flow in all three types of porosity (Darcyan flow in porous media, laminar flow in fractures, and turbulent flow in conduits) and to account for the exchange of flow between these media. For practical purposes, various simplifications are made with a double (flow in both fracture and matrix porosity) [3, 31, 37] or triple porosity models (flow in matrix, fracture and conduit porosity) [39, 40]).

Porosity of these elements is changing over time. This means that conduit porosity can evolve from fractures and the ratio between both is changing with the maturity of the karst [5]. In different karst areas, porosity varies from 0.1 to more than 50% and is strongly dependent on the type of karst rock and the methodology used during measurements taking [5].

1.4   Zonation of a Karst Aquifer

The water infiltrating from the surface moves gravitationally downward, until it reaches the level at which all the pores in the rock are filled with water (Fig. 1.1). The surface that separates the water-saturated zone from the zone with air-containing pores is the **water table** (in case of unconfined aquifer this is also **potentiometric** or **piezometric surface**) [10, 28, 38]. The zone above the water table is the **vadose zone** (also **unsaturated zone**). Through the vadose zone, the water drains downward gravitationally. The openings are not filled with water so there is no positive pressure [28]. Below the water table is the **phreatic zone** (also **saturated zone**). In this zone, the openings are completely filled with water. The flow is driven by the variations in hydraulic head, which is the sum of pressure and elevation head [28]. Depending on the hydrological conditions, the water table is oscillating over time. The part of vadose zone in which the water table is oscillating is called the **epiphreatic zone** (also **floodwater zone** or **intermittently saturated zone**) [2, 10, 20, 28].

The aquifer receives the water from different sources (Fig. 1.1). The water from rain and snow that falls on the karst area itself is called **autogenic recharge**. If the water accumulates on adjacent regions of relatively low permeable rocks and then flows into the karst as sinking streams, this water is called **allogenic recharge** [28].

Karst aquifers can also be classified according to their relation to the low permeable formations that bound them (Fig. 1.2) [28]. If there is no confining layer between

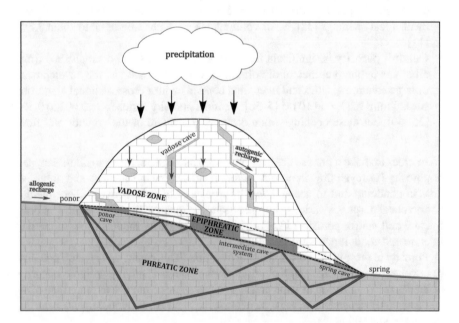

Fig. 1.1 Conceptual model of a karst hydrogeological system with its zonation (adapted from Dreybrodt et al. [8])

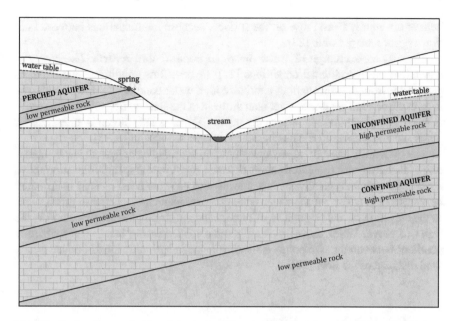

Fig. 1.2 Confined, unconfined, and perched aquifers (adapted from Dunne and Leopald [9])

the aquifer and the atmosphere, the aquifer is **unconfined** [10]. If the aquifer is positioned between two relatively impervious layers, so that it is under pressure, it is considered as a **confined aquifer** [10, 28]. If the hydraulic pressure in the confined aquifer is big enough, water can flow out of wells into the surface. This kind of aquifer is an **artesian aquifer** [28]. **Perched aquifer and perched springs** form where low permeable layers pond water above the position of the regional water table [10].

1.5 Types of Groundwater Flow in Karst Aquifers

Karst aquifers exhibit all types of flow. In matrix and fissures, **laminar flow** prevails; there, individual particles of water move in parallel threads in the direction of flow. There is no mixing or transverse component in their motion. In conduits, however, the flow is turbulent, with fluctuating eddies and transverse mixing. The transition from a laminar to a turbulent regime can be roughly assessed by the Reynolds number, which presents the ratio between the inertial and viscous forces.

Flow in conduits can be either pressurised (**pipe or closed-conduit flow**) or can have a free surface (**open-channel flow**). In the former, the cross sectional area of the flow is well defined by the conduit cross section and does not change with the rate of discharge [28]. In the case of an **open-channel flow** [28, 29], the water surface is in contact with air and changes with the amount of flow. Underground streams behave differently from those on the surface, because the transitions between both flow regimes are possible [28].

There is an **exchange of water between conduit and matrix**. Its activity is dependent on hydrological conditions [27]. In base flow conditions, the conduits gain water from the surrounding matrix. In high water conditions during floods, the head in the conduit becomes larger than the head of the surrounding matrix. The result is water flow from the conduits to the matrix. Water is then stored in intergranular pores and fractures until the head gradient is reversed again [26, 35].

The **hydraulic head** (h) at any given point in a groundwater system is a combination of elevation head (potential energy of water, positioned at specific elevation above a reference datum, typically mean sea level) and pressure head (energy contributed by pressure) [21, 28]. At the water table or in an open channel stream this is close to the elevation of that point. The difference in head from place to place makes the water to move. Water moves in the direction of decreasing head. The **hydraulic gradient** between the two points is the difference in hydraulic heads divided by the total distance of the flow [28].

1.6   The Epiphreatic Flow

Many caves with active water flow, or caves reaching the phreatic surface, are part of the epiphreatic zone. In most mature karst aquifers, such as the one studied in this work, the epiphreatic zone transfers most of the flow in medium to high flow conditions. The geometry of the epiphreatic zone is partially known from cave surveys, and partially unknown. In complex tectonic settings, such as the ones in Dinaric karst, the geometry is very complex, with a high variety of conduit diameters, termination of conduits by breakdowns, large voids related to fractured zones, and different levels of conduits. Such high irregularity of the system, together with variable recharge, results in large variations of water levels, back-flooding, overflowing, filling and emptying of large underground reservoirs, etc. The epiphreatic zone exhibits a wide variety of flow, but turbulent flow in conduits and channels prevails. The transition from full-pipe to open channel flow is frequent and has an important consequence on large-scale back-flooding [13].

When discussing the dynamics of groundwater flow in karst aquifers, the recharge conditions need to be considered. Recharge of the system can be manifold [28], which is also the case for the system discussed in this work, where a dominant part of the recharge arrives from a karst polje. In this case, many factors such as the recharge to the polje, the polje's ponor capacities, the drainage capacity and the head of the system, and the storage capacity of the polje, define the recharge to the system (Fig. 1.3).

Despite a complexity of settings, a concentrated input, such as recharge from the polje, bring into the karst characteristic transient and/or periodic signals, which, sometimes, can be followed throughout the system. Two of these parameters are briefly discussed in the following sections.

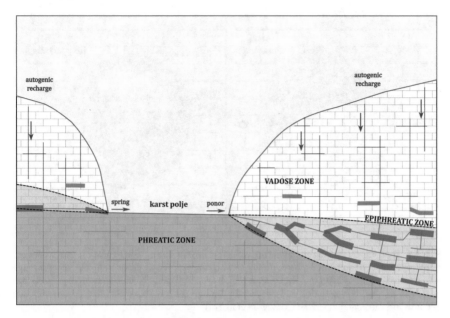

Fig. 1.3 Distribution of various wide fractures and channels in karst hydrologic zones. Blue lines denote higher density of channels in the epiphreatic zone

1.6.1 Temperature of Underground Flow

In surface streams, alternation of solar radiation and night cooling causes a diurnal oscillation of the water temperature. As water enters an aquifer, it usually has a different temperature than the aquifer. Therefore, heat exchange between water and rock occurs (Fig. 1.4) [6, 25]. The result is both damping (decrease in signal amplitude) and retardation (time lag of the signal) of the water temperature until it reaches equilibrium or the water leaves the aquifer at the spring (Fig. 1.5) [24]. Within karst aquifers, the oscillations of water temperature are often transmitted over long distances before they are fully damped [7, 25]. The damping and retardation of thermal peaks in conduits or fractures depend on the flow path's hydraulic diameter, flow-through (transit) time, and the timescale of the temperature variation. Damping and retardation also depend on the thermal conductivity, specific heat, and density of rock and the specific heat and density of water [25]. Heat exchange mechanisms in open channels might be even more complex as shown in Fig. 1.4. Covington et al. [7] derived a rule-of-thumb relation that states that the time scale of temperature variations and the underground transit time are comparable until these variations can be observed.

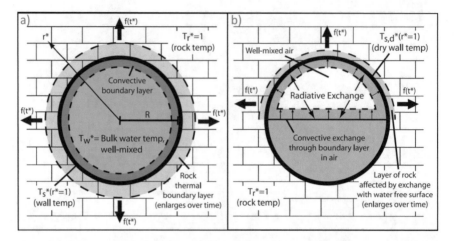

Fig. 1.4 Heat exchange in a full pipe or open channel flow: **a** The model for heat exchange between a full pipe karst conduit and the surrounding rock. Heat passes from the bulk, mixed water through a convective boundary layer into the conduit wall. **b** Water in open-channel karst conduits exchange heat with the rock via radiation and convection through the air, in addition to the processes that occur in full-pipe conduits (from Covington et al. [6])

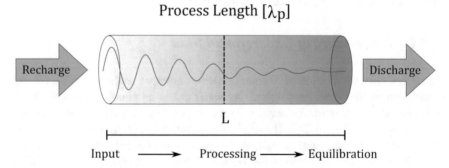

Fig. 1.5 Process length scale. Quantities input at the recharge point into a conduit (e.g., temperature, electrical conductivity) will be modified as they are transported along the conduit length, L. This modification occusrs over a characteristic length scale, λp, which is called the process length. Shading and wave amplitude indicate the extent of modification, which can be quantified using the dimensionless process number, $\Lambda = L/\lambda$p. When $\Lambda \ll 1$, little modification occurs. When $\Lambda \gg 1$ the process reaches equilibrium before discharge (adapted from Covington et al. [7])

1.6.2 Specific Electrical Conductivity (SEC) of Underground Flow

The amount of dissolved calcium carbonate in water can be expressed as carbonate hardness and is often related to specific electrical conductivity (SEC in micro-Siemens per centimetre [μS/cm]) [23, 38]. In karst waters, most ions are contributed from the H_2O-CO_2-$CaCO_3$ system. In relatively clean karst waters, the conductivity is proportional to the amount of dissolved calcite [23]. Generally, other ions, such as $K^+, Na^+, Cl^-, NO_3^-, SO_4^{2-}$, may contribute to SEC. In karst water, the range of SEC is between 150 and 1000 μS/cm [23]. Factors influencing the SEC are the residence time of water in carbonate rock, the ratio between the surface of water-rock contact and water volume, and factors defining solubility of calcite in water (e.g. presence of organic soils). Waters that enter a karst aquifer may have a different origin and composition. Unpolluted allogenic waters from nonsoluble rocks have low SEC, while water infiltrating through the epikarst and vadose zone has high SEC. During a flood event, the initial pressure pulse pushes stored *old* saturated water to conduits, which results in an initial rise of SEC at springs (and in active caves). This rise is followed by the drop of SEC caused by the arrival of relatively *fresh* floodwater. The conductivity signal is transferred along the groundwater flow and can thus be used as a natural tracer.

At various hydrological conditions, water from different sources appears in the system. Some of the water comes from karstified areas, some from non-karstified areas, and the rest comes directly from rain. Because of this, water flows through different environments and, therefore, has a different chemical composition.

During high water events, rain water with less dissolved calcium carbonate enters the aquifer. The increased recharge produces a pressure pulse that flushes out the water stored in the diffuse part of the aquifer (Fig. 1.6) [38]. This water contains more dissolved calcium carbonate, which results in increased SEC at the spring. When stored water is flushed out from the system, more dilute water flows through the system. This results in decreased SEC at the spring, which lasts until the end of the high water event (Fig. 1.6) [38].

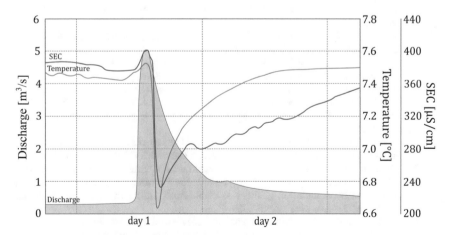

Fig. 1.6 Temporal evolution of discharge, temperature, and electrical conductivity at a karst spring in France (adapted from Tissot and Tresse [33])

References

1. Appelo CAJ, Postma D (2005) Geochemistry, groundwater and pollution. A. A. Balkema Publishers, Leiden, p 634
2. Audra P (1994) Karst Alpins - Genése de grands réseaux souterrains. Karstologia Mémoires, 5, p 280
3. Birk S, Liedl R, Sauter M (2002) Integrated approach to characterise the geometry of karst conduit systems at the catchment scale. In: Proceedings of 3rd international conference on water resources and environment research (ICWRER), Dresden, Germany, 22–26 July 2002, vol I, pp 196–201
4. Brečko Grubar V, Plut D (2001) Kakovost virov pitne vode v Sloveniji. Ujma 14–15:238–244
5. Brenčič M (1996) Konceptualni model razvoja krasa. Geologija 37–38:391–414. https://doi.org/10.5474/geologija.1995.015
6. Covington MD, Luhmann AJ, Gabrovšek F, Saar MO, Wicks CM (2011) Mechanisms of heat exchange between water and rock in karst conduits. Water Resour Res 47:W10514. https://doi.org/10.1029/2011WR010683
7. Covington MD, Luhmann AJ, Wicks CM, Saar MO (2012) Process length scales and longitudinal damping in karst conduits. J Geophys Res 117:F01025. https://doi.org/10.1029/2011JF002212
8. Dreybrodt W, Gabrovšek F, Romanov D (2005) Processes of speleogenesis: a modeling approach. Inštitut za raziskovanje krasa ZRC SAZU, Postojna, p 375
9. Dunne TR, Leopold LB (1978) Water in environmental planning. Freeman, San Francisco, p 818
10. Ford DC, Williams PW (1989) Karst geomorphology and hydrogeology. Academic Division of Unwin Hyman Ltd, London, p 601
11. Ford DC, Williams PW (2007) Karst hydrogeology and geomorphology. Willey, Chichester, p 562
12. Frantar P (ed) (2008) Water balance of Slovenia 1971–2000. Ministrstvo za okolje in prostor, Agencija Republike Slovenije za okolje, Ljubljana, p 119
13. Gabrovšek F, Peric B, Kaufmann G (2018) Hydraulics of epiphreatic flow of a karst aquifer. J Hydrol 560:56–74. https://doi.org/10.1016/j.jhydrol.2018.03.019
14. Gams I (2004) Kras v Sloveniji v prostoru in času. Inštitut za raziskovanje krasa ZRC SAZU, Postojna, p 515
15. Goldscheider N, Drew D (eds) (2007) Methods in Karst hydrogeology. Taylor & Francis, Leiden, p 264
16. Goldscheider N, Chen N (2017) World Karst Aquifer Map 1:40,000. BGR, IAH, KIT & UNESCO, Berlin, Reading, Karlsruhe & Paris
17. Gospodarič R, Habič P (eds) (1976) Underground water tracing: investigations in Slovenia 1972–1975. Inštitut za raziskovanje krasa ZRC SAZU, Postojna, p 312
18. Gostinčar P (2016) Geomorphological characteristics of karst on contact between limestone and dolomite in Slovenia. University of Nova Gorica, Graduate School, Nova Gorica, p 276
19. Gvozdetski NA (1967) Occurrence of karst phenomena on the globe and problems of their typology. Earth Res 7:98–127
20. Häuselmann P, Jeannin PY, Lauritzen SE, Monbaron M (2002) The role of the epiphreatic zone and the surrounding environment in cave genesis. In: Gabrovšek F (ed) Evolution of the karst: from prekarst to cessation. Inštitut za raziskovanje krasa ZRC SAZU, Postojna, pp 309–318
21. InTeGrate—Interisciplinary Teaching about Earth for a Sustainable Future (2018) Potential energy and hydraulic head. [Online] Available from: https://www.e-education.psu.edu/earth111/node/931. Accessed 7 April 2020
22. Kidd K, Hufman G (2011) Global precipitation measurement. Meteorol Appl 18:334–353
23. Krawczyk WE, Ford DC (2006) Correlating specific conductivity with Total hardness in limestone and dolomite Karst waters. Earth Surf Processes Landforms 31:221–234. https://doi.org/10.1002/esp.1232

24. Luhmann AJ, Covington MD, Alexander SC, Chai SY, Schwartz BF, Groten JT, Alexander EC Jr (2012) Comparing conservative and nonconservative tracers in karst and using them to estimate flow path geometry. J Hydrol 448–449:201–211. https://doi.org/10.1016/j.jhydrol. 2012.04.044

25. Luhmann AJ, Covington MD, Myre JM, Perne M, Jones SW, Alexander Jr EC, Saar MO (2015) Thermal damping and retardation in karst conduits. Hydrol Earth Syst Sci 19:137–157. https://doi.org/10.5194/hess-19-137-2015

26. Martin JB, Screaton JE (2001) Exchange of matrix and conduit water with examples from the Floridan aquifer. In: Kuniansky EL (ed) U.S. Geological Survey Karst Interest Group Proceedings. Water-Resources Investigations Report 01-4011, Denver, pp 38–44

27. Mayaud C, Wagner T, Benischke R, Birk S (2016) Understanding changes in the hydrological behaviour within a karst aquifer (Lurbach system, Austria). Carbonates Evaporites 31:357–365. https://doi.org/10.1007/s13146-013-0172-3

28. Palmer AN (2007) Cave geology. Cave Book, Dayton, p 454

29. Perne M (2012) Modelling speleogenesis in transition from pressurised to free surface flow. University of Nova Gorica, Graduate School, Nova Gorica, p 76

30. Ravbar N (2007) The protection of karst waters. Inštitut za raziskovanje krasa ZRC SAZU, Postojna, p 254

31. Sauter M (1993) Double porosity models in karstified limestone aquifers: field validation and data provision. In: Gunay G, Johnson AI, Black W (eds) Hydrogeological processes in Karst terrains, proceedings of the Antalya symposium and field seminar, October 1990. IAHS Publication number 207, pp 261–279

32. Šušteršič F (2002) Where does underground Ljubljanica flow? RMZ Mater Geoenviron 49(1):61–84

33. Tissot G, Tresse P (1978) Les systèmes karstiques du Lison et du Verneau - région de Nans-sous-Sainte-Anne (Doubs). University of Franche-Comté, Besancon, p 134

34. Turk J (2010a) Hydrogeological role of large conduits in karst drainage system. University of Nova Gorica, Graduate School, Nova Gorica, p 305

35. Turk J (2010b) Dinamika podzemne vode v kraškem zaledju izvirov Ljubljanice - Dynamics of underground water in the karst catchment area of the Ljubljanica springs. Inštitut za raziskovanje krasa ZRC SAZU, Postojna, p 136

36. Vlahović I, Tišljar J, Velić I, Matičec D (2002) The Karst Dinarides are composed of relicts of a single Mesozoic platform: facts and consequences. Geol Croat 55(2):171–183. https://doi.org/10.4154/GC.2002.15

37. Warren JE, Root PJ (1963) The behavior of naturally fractured reservoirs. Soc Petrol Eng J, 245–255

38. White WB (1988) Geomorphology and hydrology of karst terrains. Oxford University Press, New York, p 464

39. Worthington SRH (1999) A comprehensive strategy for understanding flow in carbonate aquifers. In: Palmer AN, Palmer MV, Sasowsky ID (eds) Karst modelling. Karst Waters Institute, Charlottesville, VA, pp 30–37

40. Worthington SRH, Ford DC, Beddows PA (2000) Porosity and permeability enhancement in unconfined carbonate aquifers as a result of solution. In: Klimchouk AB, Ford DC, Palmer AN, Dreybrodt W (eds) Speleogenesis, evolution of karst aquifers. National Speleological Society, Huntsville, pp 77–90

Chapter 2
Study Area

The study area is the northern part of the Ljubljanica River recharge area, limited by Planinsko Polje on the south, by Logaško Polje on the west, by the elevated region of the plateau Logaški Ravnik on the east, and by the Ljubljana Basin on the north. Geomorphologically, the whole recharge area is a combination of poljes that are surrounded by higher elevated karst plateaus. The whole recharge area geologically belongs to the External Dinarides with a thrust and nappe structure. Idrija Fault Zone presents a flow barrier, which at the same time deflects general flow direction along the system of poljes aligned along the dinaric, NW-SE direction. There are two main water flows. One is coming from a predominantly impermeable part in the southwest, while the other is from a predominantly karstified part in the southwest of the recharge area. Both flows join together in the cave Planinska jama and spring as the Unica River. After flowing over Planinsko Polje, it continues its flow underground towards the springs of the Ljubljanica River in the north of the recharge area. To give a comprehensive overview of the study area, different aspects relevant for the research, ranging from a broader to a more local scale, are presented.

2.1 Geomorphological Overview of the Broader Area

The Ljubljanica River recharge area mostly belongs to the Notranjska region, located in southwestern Slovenia. Its total recharge area is estimated to be 1900 km², whereas the recharge area upstream from the springs of the Ljubljanica River is almost 1200 km² [26] (Fig. 2.1).

The relief of the Ljubljanica River recharge area is diverse. Bordered on the north by the Ljubljana Basin and on the south by the Snežnik Mountain, it is characterised by High Dinaric plateaus [25], large karst plains, and a set of Dinaric poljes. Poljes of the Ljubljanica River recharge area stretch along the Idrija Fault Zone, a major tectonic structure in the area with Dinaric (NW-SE) orientation. Poljes are considered

© The Editor(s) (if applicable) and The Author(s), under exclusive license   13
to Springer Nature Switzerland AG 2020
M. Blatnik, *Groundwater Distribution in the Recharge Area*
of Ljubljanica Springs, Springer Theses, https://doi.org/10.1007/978-3-030-48336-4_2

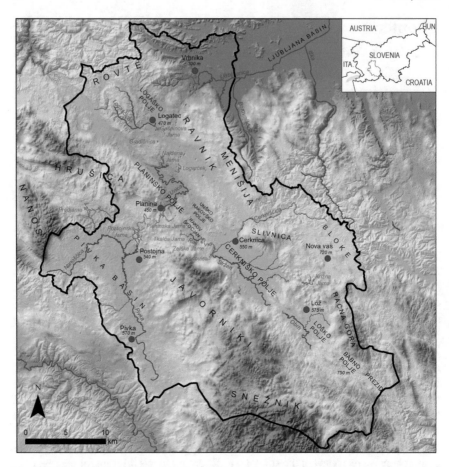

Fig. 2.1 Geographical units of the Ljubljanica River recharge area. Black line denotes the estimated recharge area of the Ljubljanica River Springs (DEM data from ARSO [4]; Cave data from Cave Register [10])

to be tectonically conditioned, but the main mechanism of their formation is corrosional planation at the piezometric level [50]. Starting from the southeast, a cascade of eight poljes can be followed towards the northwest: a polje near Prezid, Babno Polje, Loško Polje, Cerkniško Polje, Unško-Rakovško Polje, Rakov Škocjan, and Planinsko Polje. An additional two poljes that are not directly positioned within the Idrija Fault Zone are Bloško and Logaško Polje [26]. The whole set presents a unique flow system, with surface rivers on the poljes and underground flow connecting the subsequent poljes. Table 2.1 presents the main characteristics of the poljes.

Another large morphological unit is the Pivka Basin (530–550 m a.s.l.) in the southwest part of the system. With an area of 75 km^2, it is the biggest depression in the recharge area. The floor is covered with flysch [48]. The unique geomorphic and recharge characteristics define the basin as a peripheral polje [24].

Table 2.1 Some characteristics of poljes within the Ljubljanica River recharge area [23, 24, 50]

Name of polje	Elevation (in m a.s.l.)	Area (in km^2)	Type of polje
Prezid	770	~1.5	Border
Babno Polje	750	~3	Border
Bloško Polje	720	13	Border
Loško Polje	575	12	Overflow
Cerkniško Polje	550	38	Overflow + border
Unško-Rakovško Polje	520	~2	Border
Rakov Škocjan	500	~0.25	Overflow/karst window
Planinsko Polje	450	10	Overflow
Logaško Polje	475	7	Border

All depressions are surrounded by High Dinaric plateaus. Between the Pivka Basin and the set of poljes of Notranjsko podolje are the plateaus of Javorniki (up to 1268 m a.s.l.) and Snežnik (up to 1796 m a.s.l.) with about 400 km^2 of total area. Both plateaus are highly karstified with no surface stream. Pivka Basin is bordered by Nanos and Hrušica Plateau on the northwest. The drainage from these plateaus goes mainly towards the Adriatic Basin and to a smaller extent towards the Ljubljanica River (see the black line on Fig. 2.1). Several plateaus with a combination of underground and surface flow encircle Notranjsko podolje in the east. These plateaus represent the water divides between the recharge area of the Ljubljanica River springs and the Iška River and Kolpa River recharge areas [25]. Racna gora (700–1140 m a.s.l.) is karstified and is recharging karst springs on Loško Polje. Plateau Bloke (from 750 to 800 m a.s.l.) and Slivnica Mountain (1114 m a.s.l.) have surface outflows that are recharging Bloško Polje (Bloščica Stream) and Cerkniško Polje (Cerkniščica River) [26]. The plateaus Logaški Ravnik (from 500 to 800 m a.s.l.) and Menišija (from 500 to 820 m a.s.l.) are karstified and have underground outflow towards numerous springs of the Ljubljanica River [51, 54]. In the north of Logaško Polje, there is a hilly area, Rovte, rising from 500 to 800 m a.s.l. with predominantly surface outflow (streams Petkovščica, Rovtarica, Pikeljščica, Žejski potok, and Hotenka). All this water flows towards Logaško Polje and the Ljubljana Basin [7, 26].

2.2 Geological and Hydrogeological Settings

2.2.1 Tectonics—Thrusts and Nappes

The recharge area of the Ljubljanica River tectonically belongs to the External Dinarides that comprise the prevailing part of Dinaric segment of the Adriatic-Dinaric Mesozoic carbonate platform [40]. For the External Dinarides, the thrust and nappe structure is significant. Structurally the highest is the Trnovo Nappe with two accompanying lower order structures, the Hrušica Nappe and the Sovič Thrust Block. These three units are associated in the Trnovo Thrust Series. Below it lies the Snežnik Thrust Unit, which is conditionally as-sociated in the Velebit Thrust Series [38, 41].

The Trnovo Nappe is the most northwestern tectonical unit of the Ljubljanica River recharge area (Fig. 2.2). It is comprised of the northern part of the Hrušica Plateau and Rovte Plateau north of Logaško Polje. It consists of rocks of the Paleozoic and Triassic basement of the Adriatic-Dinaric Carbonate Platform, which are overlain by rocks of the carbonate platform of Upper Triassic to Upper Cretaceous rocks [34, 41]. The youngest rocks from Upper Cretaceous and Paleogene period cover the northwestern part of the nappe and dip toward the northwest [41], while the southwestern part of the carbonate platform is covered with Carboniferous-Permian clastites [40].

The Trnovo Nappe is thrusted over the **Hrušica Nappe**, located in the southeast (Fig. 2.2). It consists of Mesozoic Carbonate Platform, on which Paleogene rocks of carbonate marlstone and flysch are unconformably deposited. Within the Ljubljanica River recharge area, the Hrušica Nappe is representing the Hrušica Plateau with mountain Planinska Gora. The Sovič Thrust is a miniature pendant of the Hrušica Nappe [41] and represents the hill Sovič on north of the Pivka Basin.

The Hrušica Nappe is thrusted over the **Snežnik Thrust Unit**, which continues as the Vinodol Thrust unit and Velebit Nappe as part of the Velebit Thrust Series. It is composed of the rocks of the Adriatic-Dinaric Carbonate Platform. Predominantly Cretaceous limestones are thrusted over the Paleogene flysch sediments. The Snežnik thrust unit comprises the Pivka Basin and Javorniki and Snežnik Plateau and an extensive area southeast of it [38, 48]. Because of thrusting and further underthrusting of the Adriatic Microplate under the Dinarides, the frontal zone of the Snežnik Thrust Unit was exposed to morphological deformations. They are visible at the contact with a flysch cover of the Pivka Basin (Fig. 2.2) [37].

2.2.2   Tectonics—Faults

In the External Dinarides, three groups of faults are of the highest importance. The first group is comprised of the Ljubljana-Imot-ski fault zone [52] with the most important Želimlje Fault. The second is the fault zone of the Idrija Fault. And, the third is located more toward the southwest and contains the important Predjama, Vipava, Raša, and Divača faults (Fig. 2.2) [41].

Within the Ljubljanica River recharge area, the most important tectonic structure is the Idrija Fault Zone with a NW-SE direction (Fig. 2.2). The fault stretches from the Soča River valley towards Idrija, Kalce, and the Planinsko and Cerkniško poljes to the upper Kolpa River valley [26]. The Idrija Fault Zone is a right-lateral strike-slip with a horizontal movement supposed to be around 2.5 km [39, 56]. Its activity is still high, about 1 cm per a year [12]. The fault zone is represented by several

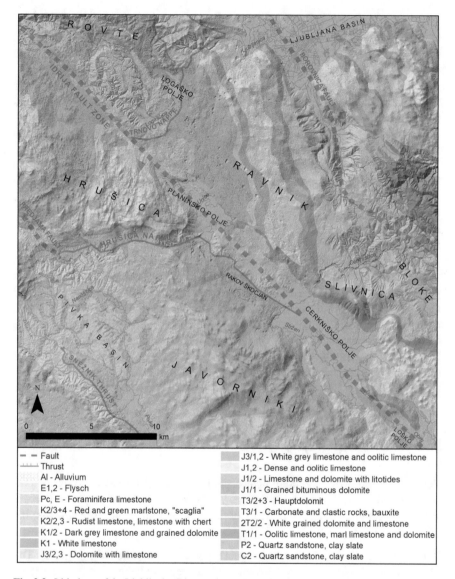

Fig. 2.2 Lithology of the Ljubljanica River recharge area with the main faults and thrusts (Osnovna geološka karta 1:100,000 [36]; DEM data from ARSO [4])

parallel faults with a total width of about 500 m [54, 56]. There are also many small transverse faults. Among these faults the Idrija Fault Zone is the youngest and the most active [13]. Within the Idrija Fault Zone, the set of poljes of the Notranjsko podolje is positioned. The geometry of the Planinsko, Cerkniško, and Loško poljes seem to be strongly influenced by tectonic activity [44]. One reason could be due to a pull-apart mechanism, which causes the releasing of the fault zones [44], but the geometry of the poljes do not completely match with the displacement and direction of the movements. However, the pull-apart mechanism is a possible process along the Idrija Fault Zone, but to a smaller extent. This is indicated by a small elongated depression in the central part of the Cerkniško Polje [56].

Except for the Idrija Fault Zone, there is no regional fault evident that crosses the area of the Ljubljanica River recharge area. The Predjama Fault passes its western border near the northern side of the Pivka Basin. The Borovnica Fault crosses north-eastern part of the Ljubljanica River recharge area over the Lubljana Basin and a small part of the Cerkniščica River recharge area (Fig. 2.2).

Besides the mentioned regional faults, the area is broken by numerous local faults that are parallel or transversal to the Idrija Fault Zone. Some examples from the area between the Planinsko and Cerkniško poljes are Grčarevski, Hotenski, Lanskovški, Babindolski, Mačkovski, Milavčev, Rakov, Koliševski, Vodendolski, Slavendolski, and Smrečnica faults [13, 14, 51]. In the area of the Pivka Basin there are Predjama, Selški, Rakulški, Sajevški, and Šembijski faults [48]. Near the springs of the Ljubljanica River, the Vrhnika Fault and some other parallel faults are located [11]. All these faults are represented by fissures, fractures, crushed and broken zones, which play an important role in the development of some of the geomorphological units and distribution of the groundwater flow. Fracture zones are favorable for development of the ponor zones, long horizontal caves, or collapse dolines [14]. Fault zones are therefore an important element that controls the groundwater flow. In some places, they act as a barrier for the groundwater flow forcing a change in the direction of the flow, while in other places they deflect the flow field along the fault orientation [26, 51].

2.2.3   Lithostratigraphy

Permo-carbonian and Permian beds occur in several belts west of Vrhnika (south-western part of the Ljubljana Basin) (Fig. 2.2). They consist of clayey and quartz sandstone, conglomerate, and partially of dolomite and limestone. The stratigraphic thickness is up to 100 m [26].

Triassic rocks are well represented, especially north of Logaško Polje (Fig. 2.2). The oldest Triassic rocks from the Schytian stage are represented by oolithic limestone, marly limestone, dolomite, and sandstone. Up to 600 m thick formations are located north of Logatec near Rovte and Zaplana. The Ladinian stage is represented by white grained dolomite and limestone, partially with cherts and tuffs. Up to 400 m thick formations are in Rovtarica and Petkovščica streams recharge area north of Logatec. Carnian stages are represented by 400 m thick formations of carbonate and clastic rocks, mostly limestones with cherts and massive dolomite. These rocks are located in the Rovtarica and Cerkniščica streams recharge area. The most represented Triassic formation is a thin bedded granular massive dolomite from the Norian and Rhaetian stages, also known as Hauptdolomit. The areas containing these rocks are around

Logaško Polje, to the east and south of the Hrušica Plateau, between Planinsko and Cerkniško poljes, and on the Bloke Plateau (Fig. 2.2). The stratigraphic thickness is between 100 and 1300 m [26, 43].

Jurassic rocks underlie areas between the Ljubljana Basin and Cerkniško Polje, the area northeast of Cerkniško Polje, and in the Hrušica Plateau (Fig. 2.2). Lower Liassic rocks are represented by well-karstified thick-grained bituminous dolomite, while Middle Liassic rocks are represented by limestone and dolomite with litiotides. Upper Liassic and Dogger stages are expressed in up to 800 m thick granular and oolithic limestone that underlies the area of most of the Ljubljanica River springs. Lower Malmian stages are represented by white grey and oolitihic limestone, while in Upper Malmian stages, coarse granular dolomite prevails. The dolomite is crushed and displaced in vertical and horizontal directions [26, 43].

Cretaceous rocks are the most widespread within the Ljubljanica River recharge area. They compose the areas between the Planinsko Polje and the Ljubljana Basin, the whole area of Javorniki and Snežnik Plateaus, and the western part of the Hrušica Plateau (Fig. 2.2). The Lower Cretaceous rocks are up to 1200 m thick and are represented by partly micritic limestone with inliers of dark grey bituminous limestone and grained dolomite. The area is well-karstified and comprises terminal ponor zones of the Planinsko Polje and some of the longest caves, respectively. Upper Cretaceous (Cenomanian and Turonian) rocks are represented by organogenic rudist limestone and limestone with chert. Senonian rocks are located only near the village Kališe (between Planinsko Polje and Logaško Polje) and are represented by red and green marlstone [26, 43].

Tertiary rocks are represented by Paleocene and Eocene rocks. Paleocene and Eocene foraminifera limestone and Dane-Kozina limestone compose the southwestern border of the Pivka Basin, while the Pivka Basin floor is underlain by Eocene Flysch. A small area with flysch also underlies the village of Kališe (Fig. 2.2) [43].

Quaternary sediments cover the floors of karst poljes and the Ljubljana Basin. Karst poljes are mostly covered by Upper Pleistocene sediments (Fig. 2.2). On Cerkniško Polje, the thickness of sediments ranges mostly from 5 to 15 m [42], while on Planinsko Polje and Unško-Rakovško Polje they are an average of 3 m [8]. Alluvial sediments mostly consist of clay, silt, and gravel. Ljubljana Basin is mostly covered by Würm sediments with a thickness up to 100 m. The southern part of the Ljubljana Basin also consists of moor sediments composed of grey clay, silt, sand, calcareous clay, peat, and black humus [26, 43].

2.2.4 Hydrogeology

According to the hydrogeological map of the Ljubljanica River recharge area [31] the rock formations are divided according to their permeability. **Highly permeable rocks** are represented by highly karstified thin bedded, bedded and non-bedded limestones and limestone conglomerate, mostly of Cretaceaus and Jurassic age. Much of plateaus Javorniki and Logaški Ravnik and areas between Loško, Bloško, and Cerkniško poljes are comprised of these rocks (Fig. 2.3). Permeability of these aquifers can be extremely high in uniformly karstified zones, but also extremely low in tectonic passive and less karstified areas.

Medium permeable rocks are represented by bedded and non-bedded dolomites and an alternation of dolomite and limestone rocks, mostly of Triassic and lower

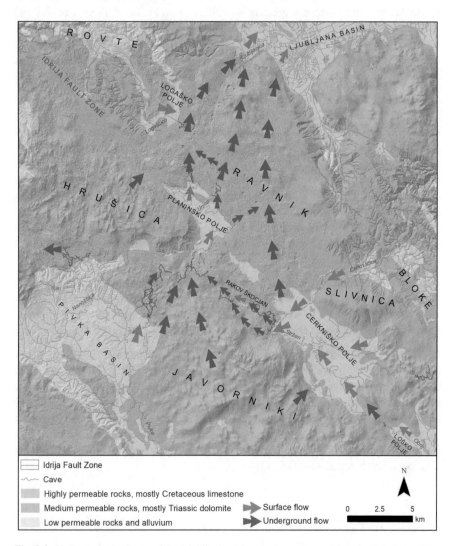

Fig. 2.3 Hydrogeological map of the Ljubljanica River recharge area with main directions of the surface and groundwater flow (modified after Krivic et al. [31]; DEM data from ARSO [4]; Cave data from Cave Register [10])

Jurassic age. In the Ljubljanica River recharge area these rocks are represented on plateaus Hrušica, Rovte, Bloke, and Slivnica (Fig. 2.3). These rocks may present barriers to limestone rocks [26].

Low permeable rocks are represented by clastic deposits with carbonate rock inliers of Permo-Carboniferous, Triassic, Cretaceous, and Paleogene age. For them fissure porosity and an unconnected underground water table is significant [27]. Impermeable layers are composed of shales, silt, and marl inlayers that present barriers for the underground water flow. Such areas are located in dispersed areas north from Logaško Polje, the whole Pivka Basin, and partially in the recharge area of the Cerkniščica River (Fig. 2.3). **Alluvium and moor sediments** are represented by clay, silt, rubble, gravel, and organic clay sediments of Quaternary age. Alluvial sediments cover all karst poljes, while moor sediments cover the southern part of the Ljubljana Basin (Fig. 2.3). The permeability of sediments quickly alters in vertical and horizontal directions [26].

On the basis of tectonical structure and lithology, several hydrogeological units can be distinguished [9, 26, 34, 41]. The most homogenous unit of the study area is the **Vrhnika-Cerknica block**, composed of highly karstified limestone and interrupted by a stripe of dolomite, the hydrogeological role of which is not clearly determined [18, 26]. According to past research several water flow pathways have developed:

- The eastern-most water flow goes directly from Cerkniško Polje towards Ljubljanica River springs. The amount of this flow is undetermined, but it surely represents a small share of the total outflow from Cerkniško Polje.
- The remaining water flows along the Idrija Fault Zone within the northern rim of Javorniki Plateau, through Rakov Škocjan towards Planinsko Polje.
- Unica River from Planinsko Polje sinks along two main ponor zones, feeding two flow paths between the polje and springs of the Ljubljanica River.

All three above mentioned pathways intertwine between Planinsko Polje and the springs of the Ljubljanica River [26, 54].

In the south is the **Rakek-Cerknica imbricated thrust**. It is composed of milonitic dolomite that presents a relative hydrogeological barrier. It is a possible reason for the groundwater flow bifurcation from the Cerkniško Polje [26]. **Snežnik-Javorniki massif** is composed of highly permeable limestone rocks. **Hrušica Plateau** is mostly composed of highly permeable limestone, but the most eastern part (mountain Planinska gora) is composed of dolomite, which presents a hydrogeological barrier [26]. Depending on the hydrological situation, the water from Hrušica Plateau flows towards Planinsko Polje or Hotenka Stream [54]. In the northwestern part of the Ljubljanica River recharge area is the **Idrija-Žiri Nappe**, which consists of the dolomite of the Idrija Nappe and the imbricated thrust of Zaplana. Because they lay on permeable limestones, they present hanging barriers [26]. Two dolomitic belts to the south of Podlipa valley and to the north of Cerkniško Polje present hydrogeological barriers, which are limiting the region of the most permeable and karstified rocks (Vrhnika-Cerknica block) and force the karst waters to rise to the surface [26].

2.3 Hydrology

The karstic recharge area of the Ljubljanica River is a highly complex system of several interacting sub-catchments composed of underground and surface flows. In this section, the main parts of these catchments are presented. Although the study is focused on the aquifer north of Planinsko Polje, the description begins with the recharge areas contributing to flow on Planinsko Polje.

2.3.1 The Area South from Planinsko Polje

In the cave Planinska jama, three waters merge to feed the springs of the Unica River and Malenšica River, the main contributors to the Planinsko Polje. These are the Pivka River (coming from Pivka basin), Rak River (coming from Notranjsko podolje), and Javorniki Stream (coming completely from Javorniki Plateau).

The Pivka River has numerous springs on the southern border of the Pivka Basin (Fig. 2.1). Its length is about 26 km [30] with a mean discharge at Prestranek gauging station of 2.65 m^3/s [17]. The Nanoščica River, with a mean discharge of 1.5 m^3/s, is the main tributary to the Pivka River [2]. During high water, the discharge of the Pivka River increases up to 55 m^3/s [2] and the Nanoščica River contributes up to an additional 22 m^3/s [2]. The Pivka Basin is also known for its numerous intermittent lakes when high piezometric level in the basin floods some of the depression. All water from the Pivka Basin sinks at its northeastern border in the cave Postojnska jama. The water continues its flow underground towards the cave Planinska jama (Fig. 2.1).

Another branch starts its flow at the southeastern part of the Ljubljanica River recharge area (Fig. 2.4) on the Croatian side. At the southern part of the polje, near the village Prezid, the stream Trbuhovica springs. Most of the year its flow is relatively short at several hundreds of meters long. During high water, its flow is extended and reaches the next Babno Polje (Fig. 2.1), where a small flooded area at its northeastern edge occurs [25]. The water springs again on Loško Polje (Fig. 2.1) from two springs, Veliki Obrh and Mali Obrh. The bigger stream, Veliki Obrh, springs at the eastern border of the polje near the village Vrhnika pri Ložu. Mali Obrh is a temporary spring, which springs in the south of the polje near the village Kozarišče. Both streams join together after about 5 km and continue their flow for the next 4 km as the river Obrh. The mean discharge is about 3 m^3/s [25], but during the highest water a large part of the polje is flooded. Most of the time, the river Obrh does not reach the polje's border because all of the water sinks in ponors in the river bed [25, 30]. During high water, it sinks at the polje's northwestern border in a 750 m long and spacious cave called Golobina, and continues its flow underground towards the Cerkniško Polje. North from Loško Polje lies Bloško Polje (Fig. 2.1) with the main water flow of Bloščica and a smaller stream Farovščica [45]. The water comes from numerous springs from the plateau Bloke that surrounds the polje.

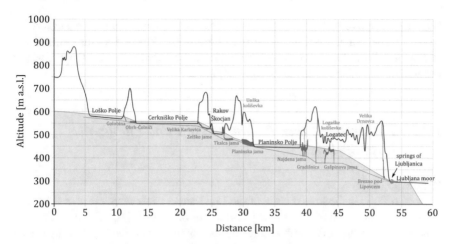

Fig. 2.4 Longitudinal cross section of the set of poljes of Notranjsko podolje with alternated surface and underground water flow [19]. It follows a broken line, initially along the NW-SE trend of the Idrija Fault Zone, and then from Planinsko Polje toward the springs of the Ljubljanica River along the northern trend. The red text denotes major caves, the cyan large collapse dolines

It sinks at the polje's western border and continues its flow underground towards the Cerkniško Polje (Figs. 2.1 and 2.4). An 8 km long cave called Križna jama is located between Bloško Polje and Cerkniško Polje, but underground water flow is not directly connected with Bloščica or Farovščica streams [45]. Cerkniško Polje is 38 km^2, is the largest polje within the Ljubljanica River recharge area, and has karstic and surface inflow [25]. The main karst spring is Jezerščica at the polje's southwestern edge. After 5 km, it joins the stream Lipsenjščica that springs at the polje's eastern border near the village of Lipsenj. Both waters continue their flow as the Stržen River. The streams Žerovniščica, Grahovščica, and Martinjščica (named after the settlements of Žerovnica, Grahovo, and Martinjak) join on the northeast side of the polje. The Stržen River flows over the western side of the polje towards a group of ponors at the polje's northwestern edge. The same ponor group is also the outflow zone for the Cerkniščica River, which comes from the northeast. It has a fluvial recharge area with springs from Bloke Plateau. The mean discharge of the Cerkniščica River is about 1 m^3/s, but during heavy rain it surpasses 50 m^3/s [2]. Floods at the polje are very common. The lowest parts of the polje are flooded most of the time—an average 286 days in a year [30]; and, at its biggest extent, an area of up to 26 km^2/s is flooded [30]. During floods, some estavelles (Rešeta, Vodonos, Retje, Zadnji kraj) are activated. During rainfall, they feed the polje, while during recession, they function as ponors. At the polje's northwestern edge, a set of ponor caves have developed. At lowest discharges, the set of ponor caves Narti and the cave Svinjska jama are activated, while during high water an 8 km long cave system, Karlovice, also drains the water [30]. Water continues its flow towards the cave Zelške Jame. The length of Zelške Jame is approximately 5.5 km and presents a spring of the river Rak [15]. This is a short surface flow over the karst window of Rakov Škocjan

(Fig. 2.1). The river gets some additional water from springs Prunkovec and Kotliči, which receive water from the Javorniki Plateau. The Rak River sinks after 2 km of flow in a 3 km long cave called Tkalca jama. The water then continues its flow towards the cave Planinska jama (Figs. 2.1 and 2.4).

2.3.2   The Area Between Planinsko Polje and Ljubljanica River Springs

Planinsko Polje
Planinsko Polje is the lowest polje of the Ljubljanica River recharge area (Fig. 2.4) with a floor elevation of 445 m a.s.l. and a size of 10 km^2 [16, 25]. It is a typical overflow-structural polje, with springs on the southern side feeding the superficial flow of the Unica River [25]. In the south of the Planinsko Polje lies a roughly 7 km long cave, Planinska jama, where there is a confluence of two water flow branches, which are described next (Fig. 2.5). From the west comes the Pivka River with the water from the Pivka Basin, and from the east comes the Rak River with the water from the set of poljes of Notranjsko podolje. A small but very stable inflow to the Rak River branch presents the water from the Javorniki Plateau [25, 30]. The water flows together as the Unica River and springs out from the spacious entrance of the cave Planinska jama (452 m a.sl.). This is an overflow spring [30] with a large range between minimum (less than 0.1 m^3/s) and maximum (90 m^3/s) discharge [2]. The mean annual discharge of the spring is about 14 m^3/s [5]. One kilometre east from Planinska jama is the spring of the Malenščica River (Malni springs) (Fig. 2.5). The springs are positioned at three levels [47] between 450 and 458 m a.s.l. It is a stable spring with a mean annual discharge of 6.7 m^3/s. The lowest discharge is 1.1 m^3/s and the highest 9.9 m^3/s. During low water, it is mostly fed by water from the Javorniki Plateau, while during high water it is also fed with water from the poljes of Notranjsko podolje and partially from the Pivka Basin [20, 30]. Because of its stable low discharge, it is used as a water supply for more than 20,000 inhabitants in the Postojna region. After about 1 km of the flow, the Malenščica River joins the Unica River. A set of smaller, temporary springs are located to the east of Malni spring. The biggest among them comes from the cave Škratovka, located close to the Hasberg gauging station (Fig. 2.5). During the strongest rain events, there is an estimated discharge of up to 7 m^3/s [30, 47]. The length of the flow is short, just about 200 m, before it joins the Unica River.

All the springs merge to the Unica River (Q_{min} = 1.1 m^3/s, Q_{mean} = 21 m^3/s, Q_{max} > 100 m^3/s [17]) and flow across the polje's surface for a total length of 17 km. For the first 7 km, the Unica River flows rather uninterrupted. When it reaches the eastern border, it starts losing water along a 2 km long reach with several groups of ponors and zones of intense leakage (Fig. 2.5). Water sinks into well expressed ponors, along lines of diffuse discharge into fractures and small dissolutional openings, and into small blind valleys entrenched into the sediment [5]. These important ponor zones are located from south to north, respectively: Mrčonovi ključi, Milavčevi

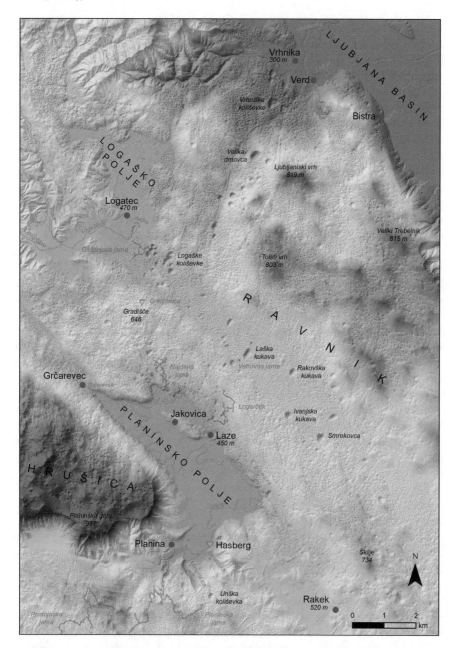

Fig. 2.5 Relief map of the studied area between the Planinsko Polje, Logaško Polje, and springs of the Ljubljanica River including the main surface waters and some of the large cave systems (DEM data from ARSO [4]; Cave data from Cave Register [10])

ključi, Ribce, Velike loke, Stara žaga pri Dolenjih lokah, and Dolenje Loke. Several studies have addressed the discharge capacity of the eastern group of ponors: 18 m³/s [28], 15.1 m³/s [26], and about 20 m³/s [51]. These estimates are close to field measurements obtained in 2015 that recorded 18.1 m³/s [5].

After 2 km of flow along the eastern border, the Unica River laterally crosses the polje and starts to follow the western border for approximately 3 km (Fig. 2.5). From the west, below the mountain Planinska gora, a set of small, higher level temporary and lower level perennial springs join to the Unica River (Fig. 2.5). Some of them were used as a water supply, but due to their small amount of discharge their importance was small [47]. On the northwestern edge of the Planinsko Polje, an important zone of estavelles called Grčarevski bruhalniki is located (Fig. 2.5). They are divided into two groups, Lebanove Rupe to the south and Brusove Rupe to the north. During strong rain events, up to 24 m³/s of water springs from numerous openings, while during flow recession, they are functioning as important ponors [46, 47].

Then, the Unica River turns northeast towards the second main group of ponors. These ponors are distributed along the northern border of the polje (Fig. 2.5). The first in the series is a ponor zone called Lebanova žaga (442 m a.s.l.), located at the foot of the hill Lanski vrh [28]. It is a terminal ponor for the Unica River during low water conditions. During higher water, it prolongs its flow towards the east and divides into two streams that lead the water towards two big groups of ponors to the north. The northernmost is a ponor zone called Pod stenami, while about 500 m to the southeast, the ponor zone Škofov lom is located. During high water, these ponors are terminal and present important outflow. According to Šušteršič [51] their outflow capacity is at least 40 m³/s.

Floods on Planinsko Polje (Fig. 2.6) occur after each intensive rain or snowmelt event [29]. The floods usually last from a few weeks to three months. According to Jenko [28] and Šušteršič [51], the Unica River starts to flood the polje when its discharge surpasses 60 m³/s at the Hasberg gauging station. In February 2014, during one of the biggest flood events recorded in the last 100 years, the water level reached 453.2 m a.s.l. and several houses in the villages of Planina and Laze were partially flooded. The lake covered an area of about 10.3 km² and nearly 80 million m³ of water was stored in the polje [1, 16].

During the beginning of the twentieth century, several attempts were made to increase the outflow capacity of ponors in order to limit the impact of flooding. Along the polje's northern side, a system of wells was constructed [46]. The wells (latter named after Putick) are a few meters in diameter and up to about 10 m deep, exposing the outflow conduits to the polje's surface. They are covered by metal bars to protect their plugging by flotsam. Along the northern and eastern borders, the ponors were also artificially widened and reinforced with concrete to improve their outflow capacity. Finally, low concrete walls were built along the eastern border to prevent outflow into the ponors during low flow conditions. During medium and high

Fig. 2.6 Partially flooded Planinsko Polje after intensive rain in September 2017 (*Photo* M. Blatnik)

water levels, the river flows over these walls into a system of channels, which guide it into the outflow points [5].

Logaško Polje
Logaško Polje is a border polje north of Planinsko Polje, developed on the contact of dolomite and limestone at an elevation between 470 and 480 m a.s.l. (Figure 2.5). It receives water from a 19 km² recharge area, where many smaller streams contribute to the Logaščica River, which sinks in the ponor zone Jačka in the town of Logatec (Fig. 2.5). The mean discharge of the Logaščica River is 0.6 m³/s, but flooding occurs when it surpasses 30 m³/s [33].

Hotenjski Ravnik is another small polje about 5 km west from Logaško Polje. It receives water from the hills on its northern side. Most of the surface streams sink along their way towards the floor of the polje. The stream Pikeljščica sinks in the cave Pucov brezen, while the stream Žejski potok sinks in the cave Kmetov brezen [47]. The only flow that reaches the floor of the polje is the Hotenka Stream. During high water, all streams reach the floor of the polje, which is partially flooded [47].

Ravnik
The plateau Logaški Ravnik is a well-karstified plain with elevated rims encircled by the Ljubljana Basin and the Borovniščica River to the north and northeast, and the set of poljes of Notranjsko podolje to the south (Fig. 2.5). The elevation is from 500 to about 800 m a.s.l. The area has a very high density of dolines, an average of about 100 per km² [6]. Besides numerous dolines, unroofed caves and collapse dolines are also common [51]. Logaški Ravnik has one of the highest densities of caves in Slovenia with more than 800 caves discovered in an area of about 130 km². Currently, 10 caves have access to the groundwater flow, which is connecting Planinsko and

Logaško Polje to the springs of the Ljubljanica River. Besides the positon of the
water caves, the distribution of collapse dolines and unroofed caves also indicate the
possible direction of the groundwater flow [51].

Springs of the Ljubljanica River
The water of the Ljubljanica River recharge area emerges at many springs located
near the town of Vrhnika, at the southern border of the Ljubljana Basin (290 m a.s.l.)
(Fig. 2.7). The line of springs generally follows the contact of Jurassic limestone and
Quaternary sediments underlain by Triassic dolomite [11] (Fig. 2.2). Over 15 springs
contribute to four main branches that join to the Ljubljanica River: Mala Ljubljanica,
Velika Ljubljanica, Ljubija, and Bistra [7] (Fig. 2.7).

Mala Ljubljanica is fed by the springs Primcov izvir, Kožuhov izvir, Mali
Močilnik, and Veliki Močilnik (Fig. 2.7). Among them, the most abundant is Veliki

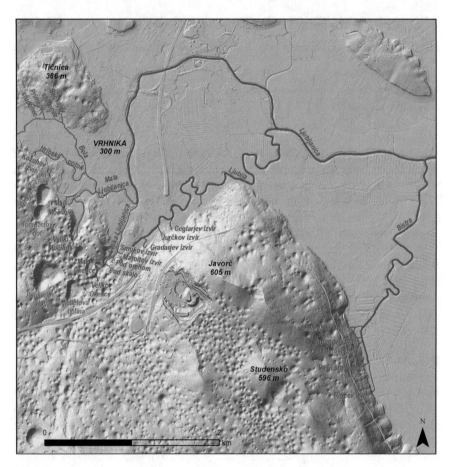

Fig. 2.7 Location of Ljubljanica River springs and collapse dolines near Vrhnika (DEM data from
ARSO [4])

Močilnik with a mean discharge of 4.95 m³/s (data from 1973–1975), while the mean discharge of all springs is 7.63 m³/s (data from 1948–1981) [2]. The minimum discharge of Mala Ljubljanica is about 0.3 m³/s, whereas maximum discharge can be up to 26 m³/s [2]. Most of the water of Mala Ljubljanica comes from Logaško Polje and the northern ponors of the Planinsko Polje [26].

The springs Malo okence, Veliko okence, Pod skalo, Pod orehom, and Maroltov contribute to Velika Ljubljanica (Fig. 2.7). The mean discharge of all the springs is 16.45 m³/s, while the minimum discharge is about 0.3 m³/s and the maximum about 70 m³/s (all data from 1948–1981) [2]. The most important spring is Pod skalo, which contributes close to half of the water. Two springs, Malo okence and Veliko okence, are overflow springs, so during low water conditions they are dry, but during high water each of them can discharge more than 10 m³/s of water. Malo Okence and Veliko Okence are mostly fed by water from Logaško Polje and the northern ponors of Planinsko Polje, while other springs in the group receive water from the eastern part of Planinsko Polje and Cerkniško Polje [26].

Ljubija consists of springs Smukov izvir, Gradarjev izvir, Jurčkov izvir, and Ceglarjev izvir (Fig. 2.7). The mean discharge of all springs is 6.81 m³/s, while the minimum discharge is about 0.4 m³/s and the maximum about 26 m³/s [2]. The most important spring in the group is Jurčkov izvir, which contributes about 2/3 of all water into Ljubija. Springs of Ljubija receive water from Logaško, Planinsko, and Cerkniško poljes [26].

The most eastern tributary of the Ljubljanica River is Bistra, positioned in Triassic dolomites. It consists of springs Grajski izviri, Zupanovi izviri, Galetovi izviri, Pasji Studenec, and Ribčev izvir (Fig. 2.7). The mean discharge of all springs is 7.48 m³/s, while the minimum discharge is about 0.9 m³/s and the maximum about 20 m³/s [2]. Almost all the water contributes to the northernmost three springs (Grajski izviri, Zupanovi izviri, Galetovi izviri) with a similar discharge in most hydrological conditions. Springs Pasji studenec and Ribčev izvir are temporary springs with a small discharge. The springs of Bistra are fed by water from Cerkniško Polje and the eastern ponors of Planinsko Polje [26].

The total mean discharge for all the springs of the Ljubljanica River, which are listed above, is 38.5 m³/s [2].

In the immediate hinterland of Mala Ljubljanica and Velika Ljubljanica springs, seven collapse dolines have developed (Fig. 2.7). Their diameter varies between 40 and 120 m, and their depth from 20 to 55 m. The floors (from 294 to 304 m a.s.l.) are just slightly higher than the floor of the Ljubljana Basin (290 m a.s.l.) and during high water conditions they can be partially flooded (for example Grogarjev dol) [19].

2.4 Studied Ponors, Caves, and Springs

As the main focus of this research concerns dynamics of groundwater, water caves were the most important element in the study as they allowed access to the groundwater. More than 800 caves are known in the study area, but only 11 of them have an access to the groundwater. For this study, 8 such caves were selected: Najdena jama, Gradišnica, Gašpinova jama, Logarček, Vetrovna jama pri Laški kukavi, Brezno pod Lipovcem, Veliko brezno v Grudnovi dolini, and Andrejevo brezno 1. Most have permanent access to the groundwater, whereas two only have access only during high floods. Four ponors and three springs were also monitored to constrain inputs and outputs from the system (Fig. 2.8 and Table 2.2). Some morphological characteristics of the studied sites are described in the next sections. In Sect. 4.1.1, other parameters of the observation points are described. In order to streamline the descriptions in the following sections, the names of observation points have been given the abbreviations listed on Fig. 2.8 and Table 2.2.

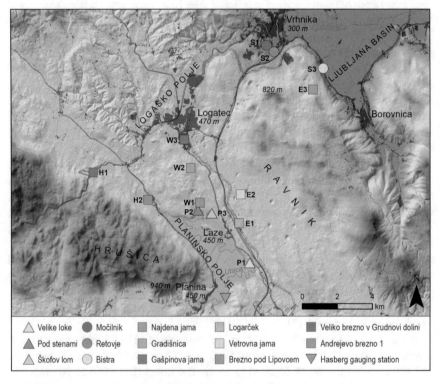

Fig. 2.8 Map of study area showing observation points at ponors (triangles), water caves (squares), springs (circles), and the official Hasberg gauging station on the Unica River (DEM data from ARSO [4]; Cave data from Cave Register [10])

Table 2.2 Some characteristics of the studied caves, ponors, springs and official gauging station [10]

Name of observation point	Length (m)	Depth (m)	Entrance elevation (m a.s.l.)	Coordinate X	Coordinate Y
P1 (ponor Velike loke)	/	/	442	444,565	78,270
P2 (ponor Požiralnik 2 pod stenami)	89	17	441	441,621	81,180
P2 (ponor Požiralnik 1 pod stenami)	37	19	441	441,641	81,175
P3 (ponor Požiralnik 1 v Škofovem lomu)	158	28	441	442,338	81,013
W1 (Najdena jama)	5216	121	518	441,800	81,619
W2 (Gradišnica)	1170	250	577	441,087	83,690
W3 (Gašpinova jama)	3375	103	482	441,080	85,450
E1 (Logarček)	4888	120	497	443,553	80,340
E2 (Vetrovna jama pri Laški kukavi)	700	114	520	444,100	82,143
E3 (Brezno pod Lipovcem)	No data	279	576	447,175	87,945
H1 (Veliko brezno v Grudnovi dolini)	110	90	495	435,605	83,315
H2 (Andrejevo brezno 1)	81	52	487	438,725	81,815
S1 (springs Močilnik)	/	/	290	445,530	90,325
S2 (springs Retovje)	/	/	289	445,955	89,972
S3 (springs Bistra)	/	/	293	448,762	89,053
Hasberg (gauging station)	/	/	445	443,185	76,294

2.4.1 Ponors and Ponor Caves

Velike loke (P1)

Velike loke is one of the ponor groups on the eastern border of Planinsko Polje (Figs. 2.5 and 2.8). The water from the Unica River is guided through the 200 m long artificially modified channel towards the ponor zone, which consists of openings and fractures. The total outflow capacity of the ponor group is 2.4 m^3/s [5]. According to the intensive tracing campaign in 1975, the water through the ponor of Velike loke mostly flows towards the springs of Velika Ljubljanica (61% of returned tracer), Ljubija (21%), and the springs of Bistra (18%). In the springs of Mala Ljubljanica, there was no returned tracer [26].

Požiralnik 2 pod stenami (P2)

Požiralnik 2 pod stenami belongs to a set of ponors on the northern rim of Planinsko Polje (Figs. 2.5 and 2.8). It is a 90 m long and 17 m deep ponor cave. The cave has two levels of passages. The upper passage has clean and smooth walls, while the lower one is covered with clay and debris and has sharp walls. During dry periods, there is no water in the cave; but during high water levels, the Unica River reaches the ponor and sinks inside (Fig. 2.9). The ponor capacity is assessed to be about 8 m^3/s (Reg. No. 98, Cave Register [10]. When the discharge of the Unica River surpasses the ponor capacity, the entire cave and also the surface above are flooded. The entrance part of the cave is walled and covered with a metal grating (Fig. 2.9), which intercepts floating debris. According to results of water tracing experiments in 1975, the water through the ponor of Požiralnik 2 pod stenami mostly flows towards the springs of Velika Ljubljanica (59%), Mala Ljubljanica (28%), and the springs of Ljubija (13%). In the springs of Bistra, there was no returned tracer [26].

Fig. 2.9   Active ponor Požiralnik 2 pod stenami (*Photo* M. Blatnik)

Požiralnik 1 pod stenami (P2)
Požiralnik 1 pod stenami (Reg. No. 97, Cave Register [10] is a short ponor cave located 30 m southeast from Požiralnik 2 pod stenami (Fig. 2.8). It is a 37 m long and 19 m deep cave that ends with water and is presumably hydraulically connected with Požiralnik 2 pod stenami. Like many ponors at this ponor zone it has an artificially walled entrance shaft covered with a metal grating.

Požiralnik 1 v Škofovem lomu (P3)
Požiralnik 1 v Škofovem lomu (Reg. No. 492, Cave Register [10] is a ponor cave that belongs to the ponor zone Škofov lom on the northern border of Planinsko Polje (Fig. 2.8). It is a predominantly horizontal cave with 158 m of length and 28 m of depth. At the lowest part, there are some accesses to the water, which present trapped water. The cave has two entrances; the upper presents a short vertical shaft, whereas the lower is an artificially walled horizontal passage representing outflow from Planinsko Polje.

2.4.2 Caves

Najdena jama (W1)
The longest cave in the system between Planinsko Polje, Logaško Polje, and the springs of the Ljubljanica River is Najdena jama with 5216 m of discovered passages (Reg. No. 259, Cave Register [10]. A cave lies at the northern border of the Planinsko Polje, with the closest chamber only 150 m away from northern ponors of the polje [26] (Fig. 2.8). The cave system has three entrances. Besides the main entrance at Najdena jama, there are also entrances at Krastača and Brezno v Galacijevem talu. Cave passages are mostly horizontal and positioned in three main levels. The upper is dry, the middle is periodically flooded, and the lowest is phreatic [49]. The groundwater level is at about 405 m a.s.l. Most of the cave was explored in the 1960s, with some important findings in the 1990s and 2000s.

Gradišnica (W2)
Gradišnica is a predominantly vertical cave about 2 km north from the cave Najdena jama and 2 km south from the town of Logatec (Fig. 2.8). It has an impressive entrance shaft with 25 m of diameter and 80 m of depth. The cave continues with the debris slope (passage Krausov hodnik) and two big chambers, the upper is Hauerjeva dvorana and the lower is Putickova dvorana (Reg. No. 86, Cave Register [10]. The lowest point presents a sump that continues with a water filled system of passages, discovered up to 50 m deep. During dry periods, the groundwater level is at 377 m a.s.l. (200 m deep in respect to cave entrance), but during high waters, it can increase for more than 50 m. During those periods, the water flows with a south-north direction [35, 53].

Gašpinova jama (W3)

The third longest cave in the studied area is Gašpinova jama with a total length of 3375 m (Reg. No. 8000, Cave Register [10]. The entrance lies in the southeastern part of the town of Logatec and stretches below the settlement (Fig. 2.8). Except for the entrance series of small shafts, the whole cave is predominantly horizontal. It is characterised by large passages and breakdown chambers with perched lakes. One passage leads towards the northwest, where the border of Logaško Polje and the ponor group Jačka are located. Another passage leads towards the southeast and ends with a sump (Fig. 2.10). During dry periods, the groundwater level is at 374 m a.s.l., but it can rise for more than 50 m. During that time, most of the passages are flooded [53].

Logarček (E1)

The entrance of the cave Logarček is located northeast of the village Laze and about 400 m away from the border of the Planinsko Polje (Fig. 2.8). With 4888 m of length, it is the second longest cave in the studied area (Reg. No. 28, Cave Register [10]. The cave is 120 m deep and has three levels of passages. The upper one is dry, the middle one is periodically flooded, while the lowest one is always below the water level [22, 32]. The main passage comes from the southwest and continues towards the northeast and north, where it ends with a sump. During dry periods, the groundwater level is at about 408 m a.s.l. [32], but the lowest discovered point of the cave is at 379 m a.s.l.

Fig. 2.10 End sump of southeast passage of Gašpinova jama (*Photo* U. Kunaver)

Vetrovna jama pri Laški kukavi (E2)

About 2 km northeast from the border of Planinsko Polje lies Vetrovna jama pri Laški kukavi (Fig. 2.8). The first part of the cave is predominantly vertical, but then it continues with a spacious horizontal passage. The southwestern part of the passage begins with the inflow sump, while the northeastern part of the passage ends with the outflow sump. Along the whole passage, an active water stream is present during all hydrological situations. The groundwater level during base flow is at 405 m a.s.l., but during high water it increases for about 25 m. The northeast part is located just below the floor of the collapse doline Laška kukava (Reg. No. 8167, Cave Register [10].

Brezno pod Lipovcem (E3)

The only periodically flooded cave that is located close to the springs of the Ljubljanica River is Brezno pod Lipovcem. It lies about 1 km southwest from the spring of Bistra on the slope of the hill Lipovec (Fig. 2.8). With a depth of about 279 m, it is the second deepest cave in the Ljubljanica River recharge area. There is no direct access to the groundwater level, but during high water events, the lowest 8 m of cave can be flooded.

Veliko brezno v Grudnovi dolini (H1)

The cave Veliko brezno v Grudnovi dolini is one of two observed caves located south from the Idrija Fault Zone. It is located southwest from the settlement of Kalce on the slope of the Hrušica Plateau, which recharges the cave [26] (Fig. 2.8). The cave is predominantly vertical with a total depth of 90 m and groundwater level at 436 m (Reg. No. 2182, Cave Register [10]. During strong rain events, it has relatively quick dynamics of the water level increase. During the highest events the whole cave can be filled, so it can function as a spring [21].

Andrejevo brezno 1 (H2)

Andrejevo brezno 1 lies next to the settlement of Grčarevec and is one of two observed caves located south from the Idrija Fault Zone (Fig. 2.8). It is a predominantly vertical cave with a depth of 52 m (Reg. No. 9641, Cave Register [10]. There is no direct access to the groundwater, but during high water events, up to 31 m of the cave can be flooded.

2.4.3 Springs

Močilnik—Veliki Močilnik (S1)

One of the biggest of the Ljubljanica River springs is Veliki Močilnik (part of the group Mala Ljubljanica springs) (Figs. 2.7 and 2.8). The mean discharge is about 4.95 m³/s, while the lowest discharge is 0.1 m³/s and the highest 20 m³/s (data from 1973–1975) [2]. The elevation of the spring is at 289.3 m a.s.l. Water tracing tests from 1975 showed that the spring of Veliki Močilnik receives water from rivers Pikeljščica,

Fig. 2.11 One of the springs of the Ljubljanica River—Izvir pod skalo (*Photo* M. Blatnik)

Žejski potok, Hotenka, Logaščica, and the northern ponors of the Planinsko Polje [26].

Retovje—Izvir pod skalo (S2)

Izvir pod skalo (Fig. 2.11) spring is part of the group Retovje that belongs to the springs of Velika Ljubljanica (Figs. 2.7 and 2.8). With a mean discharge of 8.8 m³/s, it is the most abundant [26]. The elevation of the spring is at 290.8 m a.s.l. Tracing tests from 1975 showed that Izvir pod skalo spring receives water from Logaško Polje (Logaščica River ponor), from all the ponors of the Unica River on the Planinsko Polje, and also from the ponor Vodonos on the Cerkniško Polje [26].

Bistra—Galetovi izviri (S3)

Galetovi izviri is a group of Bistra springs, which joins the Ljubljanica River after 3 km of flow (Figs. 2.7 and 2.8). The mean discharge is 2.6 m³/s, which is about 1/3 of the mean discharge of the whole Bistra. The elevation of the spring is at 292.6 m a.s.l. According to tracing tests, the springs Galetovi izviri receive the water from the southernmost ponors of Planinsko Polje (ponors Ribce and Milavčevi ključi) and from the ponor Vodonos on Cerkniško Polje [26].

Hasberg

Hasberg is an official hydrological station at the Unica River, where the Slovenian Environment Agency is operating [2] (Fig. 2.8 and Table 2.2). It is situated on the southern part of Planinsko Polje next to the remnants of the Hasberg mansion. The elevation of the station is 445 m a.s.l. The measuring parameters are temperature, water level, and discharge, measurements of which date back to 1926.

2.5 Climate of the Ljubljanica River Recharge Area

The studied groundwater dynamics are highly dependent on the climate character-istics of the recharge area. The whole recharge area of the Ljubljanica River has a transient continental climate that has the influence of both submediterranean and continental climates. The mean annual temperature in the period between 1981 and 2010 varied from about 3 °C on the higher elevated Dinaric plateaus to about 10 °C in the Ljubljana Basin [3, 55]. The mean annual precipitation in the period between 1981 and 2010 was from about 1400 mm in the Ljubljana Basin in the north and the Pivka Basin in the south to about 2500 mm on the high Dinaric plateaus in the southeast and northwest. The average of the whole recharge area was from 1600 to 1800 mm. The rain was relatively uniform in its distribution, but more wet periods usually occurred in late spring and late autumn [3, 55] (Fig. 2.12).

During the observation period (from January 2015 to May 2018) the Ljubljanica River recharge area, on average, received about 5000 mm of rain, which is about 1450 mm of rain annually [3]. During the year, the rain is relatively well-distributed, but spring and autumn months usually receive more rain. With a combination of snow melt during the spring months and lower evapotranspiration in autumn, these periods are more prone to the occurrence of high water events.

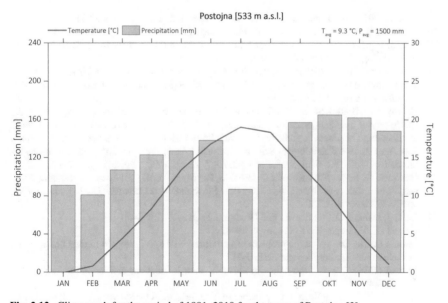

Fig. 2.12 Climograph for the period of 1981–2010 for the town of Postojna [3]

References

1. ARSO (2014) Hidrološko poročilo o poplavah v dneh od 8. do 27. februarja 2014. Agencija Republike Slovenije za okolje, Ljubljana
2. ARSO (2018a) Hydrological archive. [Online] Available from: http://vode.arso.gov.si/hid arhiv/. Accessed 7 April 2020
3. ARSO (2018b) Meteorological archive. [Online] Available from: http://meteo.arso.gov.si/met/ sl/archive/. Accessed 7 April 2020
4. ARSO (2018c) Lidar data fishnet. [Online] Available from: http://gis.arso.gov.si/. Accessed 7 April 2020
5. Blatnik M, Frantar P, Kosec D, Gabrovšek F (2017) Measurements of the outflow along the eastern border of Planinsko Polje, Slovenia. Acta Carsologica 46(1):83–93. https://doi.org/10. 3986/ac.v46i1.4774
6. Breg Valjavec M (2013) Nekdanja odlagališča odpadkov v vrtačah in gramoznicah. Geografija Slovenije, 26, Založba ZRC, Ljubljana, p 118. https://doi.org/10.13140/2.1.4860.8001
7. Brenčič M (2000) Hidrogeološka analiza velikih kraških izvirov v Sloveniji. Univerza v Ljubljani, Naravoslovnotehniška Fakulteta, Oddelek za geologijo, Ljubljana, p 357
8. Breznik M (1961) Akumulacija na Cerkniškem in Planinskem polju. Geologija 7:119–149
9. Buser S (1965) Geološka zgradba južnega dela Ljubljanskega barja in njegovega obrobja. Geologija 8:34–57
10. Cave Register (2018) Cave Register of the Karst Research Institute ZRC SAZU and Speleological Association of Slovenia. Postojna, Ljubljana
11. Celarc B, Jež J, Novak M, Gale L (2013) Geološka karta Ljubljanskega barja 1:25,000. Geološki zavod Slovenije, Ljubljana
12. Čar J, Pišljar M (1993) Presek Idrijskega preloma in potek doline Učje glede na prelomne strukture. Rudarsko metalurški zbornik 40:79–91
13. Čar J, Gospodarič R (1983) O geologiji krasa med Postojno, Planino in Cerknico. Acta Carsologica 13:91–106
14. Čar J (1981) Geološka zgradba požiralnega obrobja Planinskega polja. Acta Carsologica 10:75–106
15. Ferk M (2011) Morfogeneza kotline Rakov Škocjan. Geografski vestnik 83(1):11–25
16. Frantar P, Ulaga F (2015) Visoke vode Planinskega polja leta 2014. Ujma 29:66–73
17. Frantar P (ed) (2008) Water balance of Slovenia 1971–2000. Ministrstvo za okolje in prostor, Agencija Republike Slovenije za okolje, Ljubljana, p 119
18. Gabrovšek F, Turk J (2010) Observations of stage and temperature dynamics in the epiphreatic caves within the catchment area of the Ljubljanica River (Slovenia). Geol Croat 63(2):187–193. https://doi.org/10.4154/gc.2010.16
19. Gabrovšek F, Blatnik M (2017) Whole-day field trip (D): karst hydrology, geomorphology and speleology of the Ljubljanica Recharge area. In: Gostinčar P (ed) Milestones and challenges in karstology: abstracts & guide book = Mejniki in izzivi v krasoslovju: povzetki & vodnik, 25th International Karstological School "Classical Karst", Postojna, 2017 = 25. mednarodna krasoslovna šola "Klasični kras", Postojna, 2017. Založba ZRC, Postojna, pp 97–108
20. Gabrovšek F, Kogovšek J, Kovačič G, Petrič M, Ravbar N, Turk J (2010) Recent results of tracer tests in the catchment of the Unica River (SW Slovenia). Acta Carsol 39(1):27–37. https://doi.org/10.3986/ac.v39i1.110
21. Gams I, Habič P (1961) Brezno pod Grudnom. Proteus 24(2):58–60
22. Gams I (1963) Logarček. Acta Carsol 3:5–82
23. Gams I (1978) The polje: the problem of definition: with special regard to the Dinaric karst. Zeitschrift für Geomorphologie 22:170–181
24. Gams I (1994) Types of the poljes in Slovenia, their inundations and land use. Acta Carsol 23:285–302
25. Gams I (2004) Kras v Sloveniji v prostoru in času. Inštitut za raziskovanje krasa ZRC SAZU, Postojna, p 515

26. Gospodarič R, Habič P (eds) (1976) Underground water tracing: Investigations in Slovenia 1972–1975. Inštitut za raziskovanje krasa ZRC SAZU, Postojna, p 312
27. Habič P (1984) Vodna gladina v Notranjskem in Primorskem krasu Slovenije. Acta Carsol 13:37–78
28. Jenko F (1959) Hidrogeologija in vodno gospodarstvo krasa. Državna založba Slovenije, Ljubljana, p 237
29. Kovačič G, Ravbar N (2010) Extreme hydrological events in karst areas of Slovenia, the case of the Unica River basin. Geodin Acta 23(1–3):89–100. https://doi.org/10.3166/ga.23
30. Kovačič G (2011) Kraški izvir Malenščica in njegovo zaledje: hidrološka študija s poudarkom na analizi časovnih vrst. Univerza na Primorskem, Znanstveno-raziskovalno središče, Univerzitetna založba Annales, Koper, p 408
31. Krivic P, Verbovšek R, Drobne F (1976) Hidrogeološka karta 1: 50,000. In: Gospodarič R, Habič P (eds) Underground water tracing: Investigations in Slovenia 1972–1975. Inštitut za raziskovanje krasa ZRC SAZU, Postojna
32. Michler I (1956) Jama Logarček. Cave Register of the Karst Research Institute ZRC SAZU, Postojna, p 13
33. Mihevc A, Prelovšek M, Zupan Hajna N (2010) Introduction to Dinaric Karst. Inštitut za raziskovanje krasa ZRC SAZU, Postojna, p 71
34. Mlakar I (1969) Krovna zgradba Idrijsko Žirovskega ozemlja. Geologija 12:5–72
35. Nagode M (2016) Prvo raziskovanje Gradišnice ali Vražje jame pri Logatcu. Osnovna šola 8 talcev, Logatec, p 47
36. Osnovna geološka karta 1:100.000, List Postojna. 1963. Zvezni geološki zavod, Beograd
37. Placer L, Jamšek P (2011) Ilirskobistriški fosilni plaz – mesto na plazu. Geologija 54(2):223–228. https://doi.org/10.5474/geologija.2011.017
38. Placer L (1981) Geološka zgradba jugozahodne Slovenije. Geologija 24(1):27–60
39. Placer L (1982) Tektonski razvoj Idrijskega rudišča. Geologija 25(1):7–94
40. Placer L (2008) Principles of the tectonic subdivision of Slovenia = Osnove tektonske razčlenitve Slovenije. Geologija 51(2):205–217. https://doi.org/10.5474/geologija.2008.021
41. Placer L, Vrabec M, Celarc B (2010) The bases for understanding of the NW Dinarides and Istria Peninsula tectonics. Geologija 53(1):55–86. https://doi.org/10.5474/geologija.2010.005
42. Pleničar M (1953) Prispevek h geologiji Cerkniškega polja—Contribution to the geology of Cerkniško polje. Geologija 1:111–119
43. Pleničar M (1970) Tolmač osnovne geološke karte 1:100,000, List Postojna. Zvezni geološki zavod, Beograd
44. Poljak M (1986) Lineaments of northwest Yugoslavia and their relationship to some field measured structural elements. In: International conference on the new basement tectonics, International Basement Tectonics Association, Salt Lake City, pp 207–213
45. Prelovšek M (2014) The hydrogeological setting of Križna jama. Mitteilungen der Kommission für Quartärforschung der Österreichischen Akademie der Wissenschaften, 21, Wien, pp 27–33
46. Putick V (1889) Die hydrologischen Geheimnisse des Karstes und seine unterirdischen Wasserläufe: auf Grundlage der neuesten hydrotechnischen Forschungen. Himmel und Erde, Berlin, p 13
47. Savnik R (1960) Hidrografsko zaledje Planinskega polja. Geografski vestnik 32:212–224
48. Šebela S (2005) Tektonske zanimivosti Pivške kotline—Tectonic sights of the Pivka basin. Acta Carsologica 34(3):566–581. https://doi.org/10.3986/ac.v34i3.254
49. Šušteršič F (1981) Morfologija in hidrografija Najdene jame. Acta Carsol 10:127–156
50. Šušteršič F (1996) Poljes and caves of Notranjska. Acta Carsol 25:251–290
51. Šušteršič F (2002) Where does underground Ljubljanica flow? RMZ Mater Geoenviron 49(1):61–84
52. Tari V (2002) Evolution of the northern and western Dinarides: a tectonostratigraphic approach. EGU Stephan Mueller Special Publication Series 1, pp 223–236
53. Turk J (2008) Hidrogeologija Gradišnice in Gašpinove jame v kraškem vodonosniku med Planinskim poljem in izviri Ljubljanice. Geologija 51(1):51–58

54. Turk J (2010b) Dinamika podzemne vode v kraškem zaledju izvirov Ljubljanice—Dynamics of underground water in the karst catchment area of the Ljubljanica springs. Inštitut za raziskovanje krasa ZRC SAZU, Postojna, p 136
55. Vertačnik G, Bertalanič R (2017) Podnebna spremenljivost Slovenije v obdobju 1961–2011. 3, Značilnosti podnebja v Sloveniji. Ministrstvo za okolje in prostor, Agencija Republike Slovenije za okolje, Ljubljana, p 197
56. Vrabec M (1994) Some thoughts on the pull-apart origin of karst poljes along the Idrija strike-slip fault zone in Slovenia. Acta Carsol 23:155–167

Chapter 3
Past Research

The Ljubljanica River recharge area has been under investigation for some time. The first descriptions were made approximately 2000 years ago; and, up until the eighteenth century, studies were mostly focused on the intermittent lake on Cerkniško Polje. Early studies of the area around Planinsko Polje began with speleological investigations of the cave Planinska jama, as well as some other caves in the surrounding area. Later studies were supplemented by geomorphological and geological descriptions related to determining the polje's development. Because of flooding, many studies were dedicated to studying the dynamics of flooding, the risk of natural hazards, and investigating the potential use of poljes for hydropower electricity production. The most current investigations involve the detailed study of groundwater dynamics and the monitoring of water quality.

3.1 Early Research of the Ljubljanica River Recharge Area

The first known descriptions of the Ljubljanica River recharge area were made about 2000 years ago with the work of the Greek geographer Strabo [36]. The text describes the lake Cerkniško jezero, indicating the lake was likely well known in ancient times [33]. Works produced in the Middle ages were mostly focused on the production of regional maps and descriptions of life in the karst area.

Like the first descriptions, the first detailed studies were also done in connection with the lake Cerkniško jezero. Wernher [47] wrote about feeding and draining the lake through possible underground channels. The description mentions the annual repetition of the lake's flooding and draining, the appearance of fish with the floodwater, the fissures through which the water sinks into the caves and supplies the polje with waters from hidden reservoirs [33]. Additional descriptions of the appearance and disappearance of streams, the functioning of underground reservoirs, and the periodic occurrence of the lake were made by Kircher [17] and Brown [2, 33].

M. Blatnik, *Groundwater Distribution in the Recharge Area of Ljubljanica Springs*, Springer Theses, https://doi.org/10.1007/978-3-030-48336-4_3

Probably the most notable study is the work of Valvasor [45], who attempted to describe the recharge and discharge mechanisms of the Cerkniško Polje. Valvasor provided a detailed map of the lake naming principal caves and sinks. The theory included underground lakes, reservoirs, rivers, and sumps. He named and described various feeders and drains and the process of overflow during heavy rain events. The descriptions were supported with schematic illustrations.

In the eighteenth century, more studies were dedicated to the lake Cerkniško jezero. Nagel [24] tried to make a simplified interpretation of Valvasor's ideas. Another comprehensive work was made by Steinberg [34]. A complex theory required a series of interconnected caves at several levels to allow water to pass from dolines to springs, as well as siphons and a cave beneath the lake to account for its seasonal dryness [33, 34]. Hacquet [15] tried to explain the mechanism of the lake in a more simple way with a comparison of the hilly surrounding to a sponge, which gives water to springs after a long delay [33]. Gruber [13] introduced the concept of groundwater for the first time. He explained that the lake Cerkniško jezero is just floodwater, which occurs when inflow surpasses outflow [33].

3.2   Early Research of the Area Between Planinsko Polje and Ljubljanica River Springs

Planinsko Polje did not draw as much attention of early researchers as Cerkniško Polje. In the work of Valvasor, only a painting of the castle Mali grad, which was located next to the entrance of Planinska jama, was included. The first known description of Planinska jama was made by Nagel [24]. In 1852, Adolf Schmidl organised for the exploration of Planinska jama. The team discovered the whole Pivka River branch and more than 2 km of passages in the Rak River branch [31–33].

The first notable work about the water dynamics of Planinsko Polje was made by Putick [27]. The work was dedicated to the outflow capacity and the possible beneficial effects of reconstructing the most important ponors on some of the poljes. Related to this work, ponor reconstructions were made on Loško Polje and Cerkniško Polje, but the most intensive work was made on ponors of the Planinsko Polje. Many ponors were widened, walled, and covered with metal grids. The most well-known are two octagonal wells called Putickove štirne (Putick's wells), located in ponor zone Pod stenami at the northern border of Planinsko Polje (Fig. 3.1). Additionally, many channels were made to guide the water from the Unica River stream towards the ponors. Putick also had an important role in cave explorations. He organized and lead technically complicated explorations in the vicinity of Planinsko Polje. In 1886, he reached the groundwater level in the mostly vertical cave Gradišnica, 200 m below the surface. In that time, he also explored the cave Logarček. During 1887 and 1888, he made significant progress in the Rak River branch in Planinska jama. Some explorations were made in cooperation with famous French explorer Edouard

Fig. 3.1 Reconstructed ponors, Putickove štirne (Putick's wells), and engineer W. Putick (from [33, p. 167])

Alfred Martel [20, 27, 33]. Another work from 1892 was dedicated to Postojnska jama and the karst window Rakov Škocjan [28].

At the end of nineteenth century, Jovan Cvijić also explored the Slovene karst. The most important of his topics was dolines and the interpretation of their genesis. His most well-known analysis and interpretation is of the doline in the railway cut near Logatec. The work of Cvijić comprised descriptions of collapsed dolines, the entrance shafts of some caves, and big depressions—poljes—within the Ljubljanica River recharge area. The descriptions include their functioning along with calculations for discharges of surface streams and the dimensions of the depressions [3].

3.3 Hydropower Engineering Studies

Early systematic investigations of the polje's behaviour were related to the possible use of poljes for hydropower electricity production. In the Ljubljanica River recharge area, there were two potential poljes where accumulation of the water and further electricity production could be made. These were Cerkniško Polje and Planinsko Polje [1, 16].

Several hundred boreholes at the base and on the slopes of Cerkniško Polje and Planinsko Polje provided information about the properties of the poljes' floors. New information was obtained about the depth and composition of the sediments. The

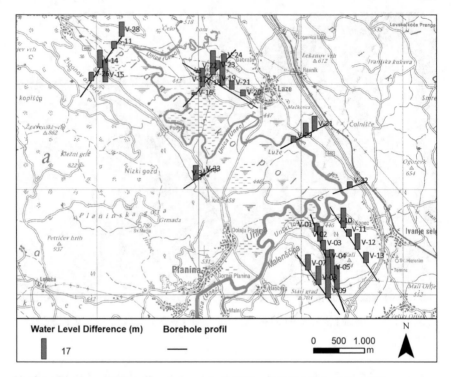

Fig. 3.2 Positions of selected boreholes made in 1950 and 1951 (obtained from [1])

thickness of the sediments on Planinsko Polje is on average 3 m, but it ranges from 1 m
to more than 20 m. The deepest points are probably characterized by ponors that are
filled with clay and are no longer active. Boreholes showed alternation of permeable
and impermeable layers of the bedrock. The measurements of groundwater level in
the boreholes showed different ranges of the fluctuation (Fig. 3.2). It ranged from
several meters below the poljes to more than 30 m at the slope of the surrounding
mountains [1, 4, 26]. Additional geophysical measurements were used to obtain more
detailed information about the depth of the sediments and the shape of the bedrock
surface [30].

Tracing experiments showed bifurcation of the water flow from Cerkniško Polje.
One part flows directly towards Ljubljanica River springs, while another flows indi-
rectly through Planinsko Polje and then towards Ljubljanica River springs [37]. Some
studies were also made in the karst caves Logarček and Gradišnica, located between
Planinsko Polje and Ljubljanica River springs. The observation showed the range of
fluctuation in each cave and provided information on which parts of the caves are
seasonally flooded [9].

The entire investigation showed that both poljes (Cerkniško and Planinsko) are
suitable for hydropower construction. The plan was to make a small accumulation
of water on Cerkniško Polje and to redirect all water towards Planinsko Polje, where

a large accumulation of water would occur. The accumulation was planned for the southwestern part of the polje, separated from all the ponor zones of the polje. The elevation of the dam would be between 470 and 480 m a.s.l and its storage more than 250 million cubic meters of water. The water would be guided towards Vrhnika through a 10 km long tube with about 80 m^3/s of discharge. In Vrhnika, there would be a hydropower plant with 120 MW of power [1]. The hydropower project in the Ljubljanica River recharge area was never realized; but, there are some in the Dinaric karst, for example on the rivers Cetina, Neretva, Trebišnjica, Zeta, and Drina [35]. Accumulations on karst poljes can present an important source of water for electricity production, but they also cause a big change in the environment. Such accumulations change the scenery, modify the habitat of protected species, and change the dynamics of the water flow. They present a potential risk for water supply contamination at an already vulnerable area [29].

3.4 Geographical Studies

Several studies were dedicated to the evolution of poljes in the past. Melik [21] was focused on the Pleistocene period. The analysis of the thickness and composition of sediments obtained from boreholes allowed for attempts at reconstructing Planinsko Polje's evolution during the Pleistocene. The discovery of lake sediments suggested that during Pleistocene the polje was flooded with a much deeper permanent lake, during which time lake sediment was deposited. Later emptying of the lake formed the current shape of the Unica River pathway and shallow dry valleys in its surrounding [21].

Gams [10] was more focused on flooding of the poljes. He conducted a study of the frequency, duration, and consequences of flooding in Planinsko Polje. Particular attention was given to the effect of artificial work. Reconstructions of ponors from the nineteenth century likely resulted in mitigating flooding, while the controlling of water in Cerkniško Polje resulted in stronger flooding events in the spring time [10].

3.5 Geological Investigations

The first basic geological map originated in 1963 with the production of a map with a resolution of 1:100,000 [25]. Later investigations focused on a smaller area and were more precise.

Čar [6] studied the geological settings of ponor zones on the border of Planinsko Polje. The goal was to identify the role of different geological structures and formations in the development of ponor zones, caves, and dolines at Planinsko Polje's floor and border. The northern and eastern part of the polje's floor is mostly underlain by well-karstified limestone, while the southern and eastern side is underlain by dolomite. The whole polje is located within the Idrija Fault Zone combined with some other, smaller fault zones, which are parallel or interconnected. Fault zones consist of fissured, crushed, and broken zones, which play a big role in the development of cavities. The ponor zones are mostly developed in fractured zones, while a smaller part is developed in voids [6].

Detailed lithological and structural mapping was conducted in the karstified area between the depressions of the Pivka Basin, Cerkniško Polje, and Planinsko Polje [5]. The study describes the reconstruction of sedimentation and tectonical processes in the past. Additionally, the role of the geological setting on karst feature development (ponors, springs, cave passages, dolines, and collapse dolines) was evaluated. Some fault zones with a SW-NE direction are more favourable for the development of young water channels in the caves Postojnska jama and Planinska jama, while faults with a NW-SE orientation are often connected with sumps and especially with the position of collapse dolines on the surface [5].

Later geological studies were dedicated to the genesis of poljes. Based on the study of the shapes of the poljes, Vrabec refuted the theory that poljes are the result of a pull-apart mechanism, as the shapes do not match with the displacement of the Idrija Fault Zone [46].

3.6 Hydrological Investigations and Tracing Tests

The first water tracing experiments in the Ljubljanica River recharge area were made in the beginning of the twentieth century. The observations confirmed an underground connection between the Pivka River and the Unica River springs. There were also confirmed connections between the Cerkniško Polje and Unica River springs with some missing information about the karst window Rakov Škocjan. The connections were confirmed between all ponors on the Planinsko Polje and two springs of the Ljubljanica River (Močilnik and Retovje). Two streams from the plateau Rovte (Petkovščica and Rovtarica) went towards the spring Kožuhov izvir near Vrhnika, while experiments on the ponors on Logaško Polje were not successful. Many tracing experiments were not successful because they were done in low flow conditions. Many other attempts are inconclusive due to the small number of observation points or a lack of precise reports about the observations [37].

Some studies of the hydrological hinterland of Planinsko Polje were made by Savnik [31]. The study contains the description of all important as well as small periodical springs in Planinsko Polje. He tried to determine if the waters from the Hotedršica region are connected to estavelles near Grčarevec. The theory posited that the situation is dependent on different hydrological conditions. Another open question was in relation to the springs of the rivers Unica and Malenščica and whether the source of their water is intertwined [22, 31].

The most comprehensive study on the Ljubljanica River recharge area was conducted in the 1970s, when extensive hydrological research had been carried out within a project related to the 3rd Symposium of Underground Water Tracing [11]. The research included numerous water tracings with some additional analyses.

Before the tracing experiment, different investigations were made including meteorological, hydrological, speleological, hydrochemical, geochemical, isotopic, bacteorological, and virological. A number of different tracers were used in the tracing (fluorescent tracers, spores, chlorides, chromium, and detergents), with which practically all important ponors between Cerkniško Polje and the Ljubljanica River springs were observed. The analysis proved numerous connections between the ponors and springs and assessed the general direction of the regional groundwater flow within the area. The results were presented in detail with precise values of concentrations and arrival times at each treated spring in addition to transit times between observed sectors. Important pathways were identified for the waters from the plateau Rovte (streams of Hotenka, Žejski potok, Pikeljščica) that flow towards the springs of the rivers Ljubljanica and Idrijca. Thus, the bifurcation between the Adriatic Sea and the Black Sea was discovered. The research identified a number of places of interweaving of the groundwater flow (Fig. 3.3). Within the system, the groundwater divides and converges [11]. Some of the most important places of interweaving in the system are the following:

- the confluence of waters from Cerkniško Polje (ponor Vodonos) and the east part of Planinsko Polje (ponors Milavčevi ključi and Ribce),
- the division of waters from Cerkniško Polje (ponor Vodonos) and the east part of Planinsko Polje (ponors Milavčevi ključi and Ribce) towards the springs that make up Bistra springs, and the spring Ceglarjev izvir,
- the confluence of waters from the north part of Planinsko Polje (ponor Pod stenami) and Logaško Polje (ponor Jačka),
- the confluence of waters from Cerkniško Polje (ponor Vodonos) and the east part of Planinsko Polje (ponors Milavčevi ključi and Ribce) with waters from the north part of Planinsko Polje (ponor Pod stenami) and Logaško Polje (ponor Jačka),
- the division of common flow in the direction towards Velika Ljubljanica and towards Mala Ljubljanica,
- the confluence of Hotenka waters (ponors of Hotenka, Pikeljščica, and Žejski potok) with the flow towards Mala Ljubljanica,
- the confluence of Rovte Stream (ponors of Rovtarica and Petkovščica) with the waters of Mala Ljubljanica (springs Kožuhov izvir and Primcov izvir),
- the bifurcation of Hotenka (ponors of Hotenka, Pikeljščica, and Žejski potok) waters towards the Ljubljanica and Idrijca rivers.

Based on the concentrations and amounts of returned tracers, the classification of springs with assessed uniform water channels was made. The classified groupings do not match totally with the groups with local names, Mala Ljubljanica, Velika Ljubljanica, Ljubija, Bistra (Fig. 3.3):

- Group 1: Mala Ljubljanica (Primcov izvir, Kožuhov izvir, Mali Močilnik, Veliki Močilnik) with part of Velika Ljubljanica (Malo okence),
- Group 2: Most of Velika Ljubljanica (Veliko okence, Pod skalo, Pod orehom, Maroltov izvir) with most of Ljubija (Gradarjev izvir, Jurčkov izvir),

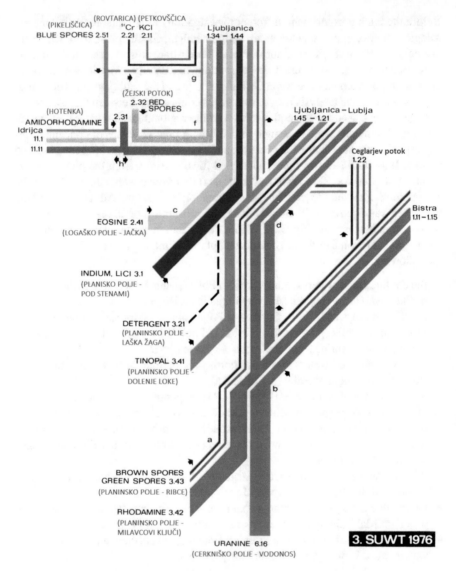

Fig. 3.3 The groundwater connections scheme according to the results of water tracing tests in 1975 (from [11], Plate XI)

- Group 3: Bistra (Grajski izviri, Zupanovi izviri, Galetovi izviri, Pasji Studenec, Ribčev izvir) with part of Ljubija (Ceglarjev izvir).

Habič [14] studied the groundwater table of the entire recharge area of the Ljubljanica River through sporadic observations of the groundwater level and flow direction in caves and springs. He determined the lowest and highest groundwater level, the gradient between selected successive locations, and presented a schematic picture of the water table at low water level (Fig. 3.4). Of course, the result is very general due to the spatial and temporal scarcity of data. The distribution of faults and different permeable formations result in separated watersheds and a gradual distribution of the karst water [14].

An additional tracing test in the Ljubljanica River recharge area was made for the vadose zone Kogovšek [18]. The study was dedicated to the velocity of percolation through the vadose zone in the area above the caves Postojnska jama, Pivka jama, and Planinska jama. The results showed very different permeable pathways with a velocity from less than 0.001 cm/s up to about 2 cm/s. The residence time through the 40–100 m deep vadose zone was mostly from 3 weeks to 3 years.

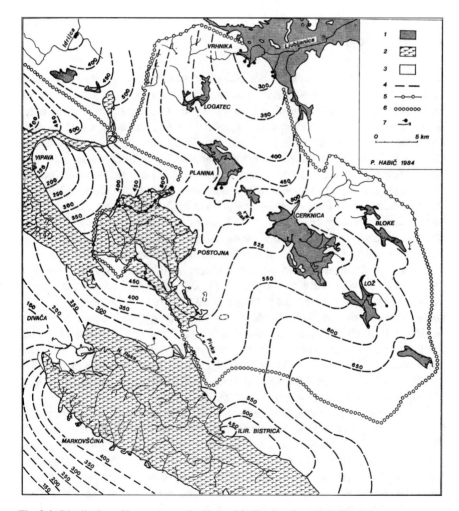

Fig. 3.4 Distribution of low waters at the Notranjska Region (from [14], Fig. 14)

One of the last tracing experiments in the Ljubljanica River recharge area was made in the Unica River hinterland. The injection points were in the ponors at the border of the Pivka Basin and Cerkniško Polje, while the observation points were within the system before the springs in the south of Planinsko Polje. Two tracing tests were made in different hydrological conditions, one during a rain event and another during the flow recession. The different conditions resulted in very different velocities of the dominant flow. During the rain event, the velocities were from 88 to 640 m/h, while during the flow recession they were 2–4 times lower. Additionally, the breakthrough curves and dispersitivity of the tracer were also dependent on the hydrological situation [8].

3.7 Speleological Studies

One of the first studies related to morphological evaluation of the caves was made by Gams [9]. The goal was to learn how the geological structure and the outflow from Planinsko Polje affect the morphology of the caves downstream from the polje. He focused particularly on the cave Logarček, while other observed caves were Mačkovica, Vranja jama, Skednena jama, and Gradišnica. Gams assessed that the shape of the passages is related to the direction and dip of the formations and also to the orientation of the fractures. Large dimensions of passages and collapses are probably the result of aggressive surface water inflow. Part of the study was dedicated to the question of why some cave waters are corrosional while others are depositional. The initial results were connected to the source of the water and the hydrological situation. Gams also discovered that cave chambers, passages, shafts, and chimneys are not necessarily related to the distribution of the dolines at the surface [9].

Another study, dedicated to the morphology of the caves near the border of Planinsko Polje, was made by Gospodarič [12]. The research analysed the distribution of horizontal and vertical passages of all known caves at Planinsko Polje's border. There are three different levels of horizontal passages within the caves, which are filled with a relatively large amount of sediments. They were developed in conditions where the relative elevation of inflow was higher than at present. The most active passages that are periodically flooded are located at the elevations from 410 to 430 m a.s.l. He assessed the position of the permanent water level in the vicinity of Planinsko Polje at about 400 m a.s.l. [12].

Šušteršič [38, 41] performed several studies in caves downstream from Planinsko Polje. According to numerous observations of the water level at different places in Najdena jama, Šušteršič pointed out three situations: During the lowest stage, the water is caught in different sumps at different levels. During moderate flooding, the water rises a few meters, but the gradient is still present. During periods of high water, the water level is at the same height in all places in the cave. Also during this time, the water level is just a few meters higher than in the Gradišnica cave 2 km to the north. A possible reason for this could be a geological barrier downstream from the cave Gradišnica, which behaves as a dam and controls the water level in the observed caves [38].

Šušteršič also presented a theoretical evaluation of the geological structure and groundwater flow between Planinsko Polje and the springs of the Ljubljanica River [41]. He anticipated that the water flowed through developed channels along well-defined corridors, which are determined by faults and lithology. Most of the channels of existing caves and unroofed caves are developed along lithological (limestone-dolomite) contacts, while sediments found in unroofed caves shows the origin of the regional flow [23, 39, 41]. The plateau Logaški Ravnik is crossed by two lineaments that are parallel to the Idrija Fault Zone. These structures delineate three corridors of the water flow, which are consistent with the results of the past tracing tests (Fig. 3.5). Two more transversal lineaments delimit the area on some individual hydrogeological blocks [40, 41].

3.8 Recent Studies and Autonomous Measurements

Modern studies are often related to autonomous measurements. The work of Kovačič [19] was comprised of comprehensive analysis of data obtained from autonomous measurements of surface streams in the recharge area of the Malenščica River spring. Automatic programmable data loggers were used for measuring the water level, discharge, and water temperature. Time series analysis was also made with data obtained from an official hydrological and meteorological station. Because of the vulnerability of the spring and its importance to the water supply, many chemical analyses and a water tracing test were also made. They proved the overlapping of the Malenščica River recharge with that of the Unica River spring, and, for the first time, indicated the possible connection with the water from the Pivka River [19].

Finally, the most recent research based on the continuous autonomous monitoring of the water was made by Turk and Gabrovšek [7, 42–44]. The study was dedicated to monitoring the groundwater flow of the Ljubljanica River in four caves between Planinsko Polje and the springs of the Ljubljanica River (Vetrovna jama pri Laški kukavi, Najdena jama, Gradišnica, and Gašpinova jama). The automatically measured parameters were water pressure (level) and temperature, which provided new insights into the dynamics of the groundwater flow in the area. The analysis of the water level hydrographs identified some regional flow directions in the caves Najdena jama, Gradišnica, and Gašpinova jama (Fig. 3.6). The combined use of the precipitation measurements showed the response to groundwater level increase to be very fast. In contrast, the duration of the recession is much slower. In the caves close to ponors, the recession is several times longer, while further in the system the inflow is damped and, therefore, the water level hydrograph is more symmetrical [7]. The stronger the rainfall event is, the higher the influence of the Unica River inflow is in comparison to the autogenic inflow [42]. In some caves (Gradišnica and Gašpinova jama), uniform groundwater level occurs. They explained that this is

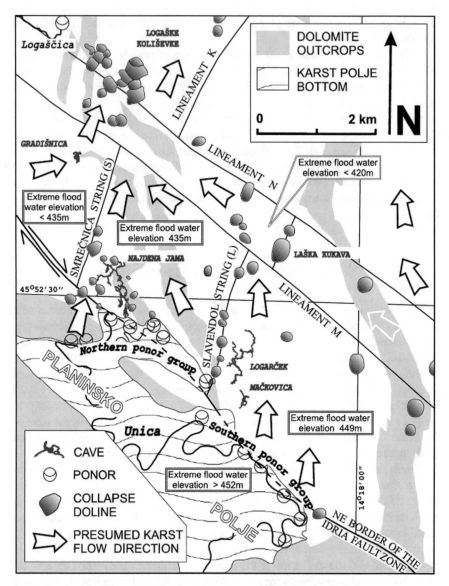

Fig. 3.5 Distribution of caves and collapse dolines and supposed main corridors of the groundwater flow (from [41, p. 72])

caused by a hydrogeological barrier between Logatec and the springs of the Ljubljanica River [7, 42, 44] (Fig. 3.6). The water level hydrograph of the cave Vetrovna jama pri Laški kukavi has a plateau-like shape. This "plateau" shape is a result of stable discharge in the cave, which is controlled by a limited outflow capacity of the eastern group of the Planinsko Polje ponors. Some small peaks above the "plateau"

Fig. 3.6 An elevation profile of the area between Planinsko Polje and the Ljubljana Basin with marked caves and estimated level and direction of groundwater flow (from [7, p. 192])

were related to the activity of higher positioned ponors [7]. Inflection points on the water level hydrographs indicate a different geometry of the system, which results in different speeds in the increase and decrease of the groundwater level [7]. Temperature records have been used to assess the transit times and flow directions between different observation points. Transit time calculations show that the velocities of the groundwater flow in the studied system differs, which is a result of different development of the channels. During high water level events, diurnal temperature oscillations also occur. Diurnal temperature oscillations in Gradišnica were recorded when the Unica River surpassed 30 m^3/s; while in the downstream cave of Gašpinova jama, oscillations only occurred with discharge far above 60 m^3/s [7]. Significantly different observed temperatures at low water conditions indicate a different origin of the inflow in the caves of Gradišnica and Gašpinova jama [42]. On the basis of all of the analyses, Turk pointed out three possible directions of the groundwater flow [44]:

- northern ponors of the Planinsko Polje—cave Najdena jama—cave Gradišnica,
- eastern ponors of the Planinsko Polje—cave Vetrovna jama pri Laški kukavi,
- eastern ponors of the Planinsko Polje—cave Gradišnica.

3.9 Shortcomings

The most useful basis for the present research is the work of Turk and Gabrovšek [7, 42–44]. It matches with the study area and also from a methodological and interpretational point of view. In spite of the good approach, some shortcomings were pointed out. During the whole observation period, only moderate flooding events were recorded. In addition, the number of observation points in the caves was relatively small (4); and, moreover, ponors and springs were not observed at all. Finally, the importance of precipitation distribution around the area was not considered in the study, nor was the usefulness of specific electrical conductivity as a potential natural tracer.

References

1. Breznik M (1961) Akumulacija na Cerkniškem in Planinskem polju. Geologija 7:119–149
2. Brown E (1673) An brief account of some travels in diverse parts of Europe ... Styria, Carinthia, Carniola ...- Tooke, London, p 144
3. Cvijić J (1898) Das Karsphänomen. Hölzl, Wien, p 113
4. Čadež N (1954) Geologija Planinskega polja in okolice. Vodnogospodarska osnova porečja Ljubljanice. Prirodne osnove. Geologija. Rokopisni elaborat projekta Nizke zgradbe, Ljubljana, p 108
5. Čar J, Gospodarič R (1983) O geologiji krasa med Postojno, Planino in Cerknico. Acta Carsologica 13:91–106
6. Čar J (1981) Geološka zgradba požiralnega obrobja Planinskega polja. Acta Carsologica 10:75–106
7. Gabrovšek F, Turk J (2010) Observations of stage and temperature dynamics in the epiphreatic caves within the catchment area of the Ljubljanica River (Slovenia). Geologica Croatica 63/2:187–193. https://doi.org/10.4154/gc.2010.16
8. Gabrovšek F, Kogovšek J, Kovačič G, Petrič M, Ravbar N, Turk J (2010) Recent results of tracer tests in the catchment of the Unica River (SW Slovenia). Acta Carsologica 39/1:27–37. https://doi.org/10.3986/ac.v39i1.110
9. Gams I (1963) Logarček. Acta Carsologica 3:5–82
10. Gams I (1980) Poplave na Planinskem polju. Geografski zbornik 20:5–35
11. Gospodarič R, Habič P (eds) (1976) Underground water tracing: investigations in Slovenia 1972–1975. Inštitut za raziskovanje krasa ZRC SAZU, Postojna, p 312
12. Gospodarič R (1981) Morfološki in geološki položaj kraških votin v ponornem obrobju Planinskega polja. Acta Carsologica 10:157–172
13. Gruber T (1781) Briefe hydrographischen und physicalischen inhalts im krain, Krauss, p 159
14. Habič P (1984) Vodna gladina v Notranjskem in Primorskem krasu Slovenije. Acta Carsologica 13:37–78
15. Hacquet B (1778–1779) Oryctographia Carniolica, oder Physikalische Erdbeschreibung des Herzogthums Krain. Breitkopf, 4 vols, Leipzig
16. Jenko F (1959) Hidrogeologija in vodno gospodarstvo krasa. Državna založba Slovenije, Ljubljana, p 237
17. Kircher A (1665) Mundus subterraneus. Jansson, 2 vols, Amsterdam
18. Kogovšek J (1997) Water tracing tests in vadose zone. In: Kranjc A (ed) Tracer hydrology, Proceeding of the 7th international symposium on water tracing, 26–31 May 1997, Portorož, Slovenia. A. A. Balkema, Rotterdam, pp 167–172
19. Kovačič G (2011) Kraški izvir Malenščica in njegovo zaledje: hidrološka študija s poudarkom na analizi časovnih vrst. Univerza na Primorskem, Znanstveno-raziskovalno središče, Univerzitetna založba Annales, Koper, p 408
20. Martel EA (1894) Les Abimes .... Delagrave, Paris, p 578
21. Melik A (1955) Kraška polja Slovenije v Pleistocenu. inštitut za geografijo, Ljubljana, p 162
22. Michler I (1955) Rakov rokav Planinske jame. Acta Carsologica 1:73–90
23. Mihevc A, Slabe T, Šebela S (1998) Denuded caves—an inherited element in the karst morphology: the case from Kras. Acta Carsologica 27/1:165–174
24. Nagel JA (1748) Beschreibung deren auf aller: höchsten Befehl ihro Röm: kayl: und königl: Maytt: Francisci I., untersuchten, in dem Herzogtum Crain befindlichen Seltenheiten der Natur. Österreichische National Bibliothek, Wien, p 97
25. Osnovna geološka karta 1:100.000, List Postojna. 1963. Zvezni geološki zavod, Beograd
26. Pleničar M (1953) Prispevek h geologiji Cerkniškega polja—Contribution to the geology of Cerkniško polje. Geologija 1:111–119
27. Putick V (1889) Die hydrologischen Geheimnisse des Karstes und seine unterirdischen Wasserläufe: auf Grundlage der neuesten hydrotechnischen Forschungen. Himmel und Erde, Berlin, p 13

28. Putick V (1892) Führer in die grotten und Höhlen, sowie in die Umgebung von Adelsberg, Lueg, Planina, St. Canzian und Zirknitz in Krain. Schäber, Postojna, p 60
29. Ravbar N (2007) The protection of karst waters. Inštitut za raziskovanje krasa ZRC SAZU, Postojna, p 254
30. Ravnik D (1976) Kameninska podlaga Planinskega polja. Geologija 19:291–315
31. Savnik R (1960) Hidrografsko zaledje Planinskega polja. Geografski vestnik 32:212–224
32. Schmidl A (1854) Die Grotten und Höhlen von Adelsberg, Lueg, Planina und Laas. W. Braumüller, 2 vols, Wien, p 316
33. Shaw T, Čuk A (2015) Slovene karst and caves in the past. Inštitut za raziskovanje krasa ZRC SAZU, Postojna, p 464
34. Steinberg FA (1758) Gründliche Nachricht von dem Innen-Crain gelegenen Czirknitzer See …. Lechner, Gratz, p 235
35. Stevanović Z, Milanović P (2015) Engineering challenges in Karst. Acta Carsologica 44/3:381–399. https://doi.org/10.3986/ac.v44i3.2963
36. Strabo (1927) Geography. Trans. H. L. Jones, Heinemann, 3 vols, London
37. Šerko A (1946) Barvanje ponikalnic v Sloveniji. Geografski vestnik 18:125–138
38. Šušteršič F (1981) Morfologija in hidrografija Najdene jame. Acta Carsologica 10:127–156
39. Šušteršič F (1998) Interaction between a cave system and the lowering karst surface case study: Laški ravnik. Acta Carsologica 27/2:115–138. https://doi.org/10.3986/ac.v27i2.506
40. Šušteršič F (2000) Speleogenesis in the Ljubljanica river drainage basin, Slovenia. In: Klimchouk AB, Ford DC, Palmer AN, Dreybrodt W (eds) Speleogenesis, evolution of karst aquifers. National speleological society, Huntsville, pp 397–406
41. Šušteršič F (2002) Where does Underground Ljubljanica Flow? RMZ Mater Geoenviron 49/1:61–84
42. Turk J (2008) Hidrogeologija Gradišnice in Gašpinove jame v kraškem vodonosniku med Planinskim poljem in izviri Ljubljanice. Geologija 51/1:51–58
43. Turk J (2010a) Hydrogeological role of large conduits in karst drainage system. University of Nova Gorica, Graduate School, Nova Gorica, p 305
44. Turk J (2010b) Dinamika podzemne vode v kraškem zaledju izvirov Ljubljanice—Dynamics of underground water in the karst catchment area of the Ljubljanica springs. Inštitut za raziskovanje krasa ZRC SAZU, Postojna, p 136
45. Valvasor JW (1789) Die Ehre dess Hertzogthums Crain …. W. M. Endter, 4 vols, Laybach & Nürnberg
46. Vrabec M (1994) Some thoughts on the pull-apart origin of karst poljes along the Idrija strike-slip fault zone in Slovenia. Acta Carsologica 23:155–167
47. Wernher G (1551) De admirandis Hungariae aqvis hypomnemation. Aquila, Wien, p 20

Chapter 4
Methods

Due to the study area's complex setting and the multiple factors and mechanisms involved in determining its groundwater dynamics, several approaches were used when conducting the research presented in the following chapter. Field work included set-up and maintenance of the autonomous measurement network, and sporadic field measurements in surface streams, caves, and springs. Data from public observation networks were also used. The data was analysed using multiple approaches. The data was interpreted through the use of basic hydraulic principles as well as incorporating geological and speleological data obtained from other works. The plausible concepts and mechanisms were tested through the use of simple hydraulic models.

4.1 Water Level, Temperature, and Conductivity Measurements

4.1.1 Autonomous Measurement Network

An autonomous measurement network was established in major ponors, caves with access to groundwater level, and springs. Instruments with autonomous loggers of pressure (depth), water temperature, and specific electrical conductivity (SEC) were installed in eight selected caves: E1 (Logarček), E2 (Vetrovna jama pri Laški kukavi), E3 (Brezno pod Lipovcem), W1 (Najdena jama), W2 (Gradišnica), W3 (Gašpinova jama), H1 (Veliko brezno v Grudnovi dolini), and H2 (Andrejevo brezno 1) (Fig. 4.1). Four measuring stations were additionally established on the ponors on Planinsko Polje: P1 (Velike Loke), P2 (Požiralnik 1 pod stenami and Požiralnik 2 Pod stenami), and P3 (Požiralnik 1 v Škofovem lomu). Three observation points were established on the springs of the Ljubljanica River: S1 (Močilnik), S2 (Retovje), and S3 (Bistra).

© The Editor(s) (if applicable) and The Author(s), under exclusive license
to Springer Nature Switzerland AG 2020
M. Blatnik, *Groundwater Distribution in the Recharge Area
of Ljubljanica Springs*, Springer Theses, https://doi.org/10.1007/978-3-030-48336-4_4

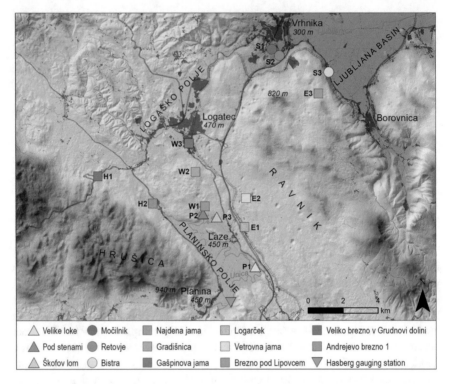

Fig. 4.1 Studied area with observation points at ponors (triangles), water caves (squares), springs (circles), and the official Hasberg gauging station on the Unica River (DEM data from [4]; Cave data from [7])

Basic characteristics of the selected ponors, caves, and springs are described in Sect. 2.4.

Several types of instruments were used. Schlumberger Diver™ instruments were used for measuring water level (pressure) and temperature. Onset Hobo U20 instruments recorded water level and temperature, while U24 units recorded specific electrical conductivity (SEC) measurements and temperature. Parameters for each instrument including range, accuracy, and resolution are listed in Table 4.1.

To calculate water depth from pressure data, the measurements were compensated with barometric data from Ljubljana and a factor of 10.2 was used to convert from pressure in bars to depth in meters.

$$Depth\ in\ m = (Total\ measured\ pressure\ in\ mbar$$
$$- Air\ pressure\ in\ mbar) * 10.2/1000$$

In the caves with groundwater access, all three parameters (water pressure, temperature, and specific electrical conductivity) were measured. There are two caves (H2—Andrejevo brezno 1 and E3—Brezno pod Lipovcem), in which the water is present

Table 4.1 Parameters for the instruments used in the study [15, 16, 21]

Parameter	Schlumberger mini diver	Schlumberger micro diver	Onset hobo U20	Onset hobo U24
Measuring parameter	Pressure, temperature	Pressure, temperature	Pressure, temperature	Conductivity, temperature
Operating pressure (in m H_2O)	10, 20, 50, 100	10, 20, 50, 100	4, 9, 30, 76	/
Pressure accuracy (in % of full scale)	±0.25	±0.25	±0.3	/
Pressure resolution (in cm H_2O)	0.2–2	0.2–2	0.1–1	/
Operating temperature (in °C)	0–50	0–50	−20 to 50	−2 to 36
Temperature accuracy (in °C)	±0.2	±0.2	±0.44	±0.1
Temperature resolution (in °C)	0.01	0.01	0.1	0.01
Operating conductivity (in μS/cm)	/	/	/	0–15,000
Conductivity accuracy (in μS/cm)	/	/	/	5
Conductivity resolution (in μS/cm)	/	/	/	1
Memory capacity (measurements)	24,000	48,000	21,700	18,500
Programmed interval (in h)	1	1	1	1

only during periods with a high water level. Therefore, only pressure and temperature were measured in these locations. This is also the case for ponors P1 (Velike loke), P2 (Požiralnik 2 pod stenami), and P3 (Požiralnik 1 v Škofovem lomu). In the three springs of the Ljubljanica River, temperature and specific electrical conductivity were measured.

Figure 4.2 shows a typical installation with U20 and U24 instruments. They are fixed into a plastic holder and attached to a stainless rod, which is attached to the wall. Installation and maintenance was performed in low-flow conditions.

Table 4.2 provides a list of caves along with their positions and types of instruments. The measurement interval was set to 1 h, while the operating time from the beginning to the end of the measurements is shown in Table 4.2 and on Fig. 4.3.

Most of the instruments for measuring water pressure were installed in January and February of 2015. Most of the instruments for measuring SEC were added later, in the summer of 2016 (Table 4.3 and Fig. 4.3). Due to the malfunctioning of some the

Fig. 4.2 Observation point with instrument with programmable data logger (*Photo* M. Blatnik)

instruments, all datasets are not complete. The available data is shown in Table 4.3 and Fig. 4.3. Manual download of the data was completed two to three times a year.

Positioning the instruments was often a challenge. An instrument must be both accessible and positioned under the water for as long as possible. The observation points that were selected for the study have different characteristics. In caves, the instruments were either positioned in lakes or sumps with no observable flow during low stage or into streams with some flow. The first type of location provides a good response to water levels, but little or no temperature response until actual flow establishes at the point due to overflows. Positions with continuous flow provide temperature records during low flow, but are, on the other hand, less responsive to level changes. Figure 4.4 shows the basic characteristics of the selected locations.

4.1.2 Manual Measurements

Besides establishing an autonomous measurement network, sporadic field measurements were also obtained. For this, a WTW Multi 3400i portable device was used (Fig. 4.5). Measured parameters were temperature (range 0–50 °C, accuracy ±0.1 °C, resolution 0.1 °C) and specific electrical conductivity (range 0–1999 μS/cm, accuracy ±0.5%, resolution 1 μS/cm). Manual measurements were taken during each visit to the observation points, while downloading the data from the automatic programmable

Table 4.2   List of observation points and their associated instrument positions (Cave data from [7])

Name of observation point	Entrance elevation (m a.s.l.)	Elevation of instrument (m a.s.l.)	X coordinate of instrument	Y coordinate of instrument
P1 (ponor Velike loke)	442	442	444,565	78,270
P2 (ponor Požiralnik 2 pod stenami)	441	430	441,625	81,197
P2 (ponor Požiralnik 1 pod stenami)	441	422	441,641	81,175
P3 (ponor Požiralnik 1 v Škofovem lomu)	441	431	442,338	81,013
W1 (Najdena jama)	518	408	441,722	81,630
W2 (Gradišnica)	577	377	441,185	83,540
W3 (Gašpinova jama)	482	374	440,815	85,110
E1 (Logarček)	497	432	443,958	80,480
E2 (Vetrovna jama pri Laški kukavi)	520	415	444,040	82,085
E3 (Brezno pod Lipovcem)	576	297	447,174	87,945
H1 (Veliko brezno v Grudnovi dolini)	495	436	435,605	83,300
H2 (Andrejevo brezno 1)	487	435	438,725	81,815
S1 (springs Močilnik)	290	290	445,530	90,325
S2 (springs Retovje)	289	289	445,955	89,972
S3 (springs Bistra)	293	293	448,762	89,053

instruments. Such measurements were necessary for cross-validation for the calibration of specific electrical conductivity (SEC). Some additional field measurements were taken during different meteorological events, for example during rain, snow melt, floods, droughts, and at normal stage. The measurements were taken in both caves and surface streams (Fig. 4.5b). These measurements provide a better coverage of measurements in various water level stages.

In order to get a clearer picture of the functioning of the eastern ponor group on Planinsko Polje, discharge measurements of the Unica River were taken (Fig. 4.5a). For this purpose, an Acoustic Doppler Current Profiler (ADCP) was used. The obtained data enabled calculations of the outflow capacity of selected ponors. These data were used for the interpretations, where the amount of water that enters the aquifer is an important factor [5].

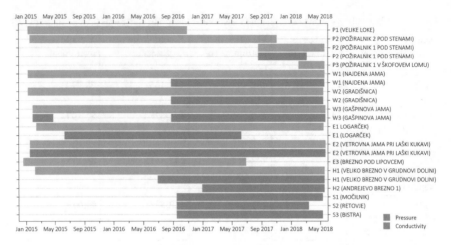

Fig. 4.3 Graph marking periods of measurement of selected parameters at selected observation points

Table 4.3 The list of observation points, types of instruments used, and period of measurement

Name of observation point	Type of instrument, range	Measuring period
P1 (ponor Velike loke)	Schlumberger, 50 m	08.01.2015–26.10.2016
P2 (ponor Požiralnik 2 pod stenami)	Schlumberger, 50 m	16.01.2015–01.11.2017
P2 (ponor Požiralnik 1 pod stenami)	Hobo, pressure, 30 m Hobo, conductivity	18.08.2017–16.05.2018 18.08.2017–04.03.2018
P3 (ponor Požiralnik 1 v Škofovem lomu)	Schlumberger, 50 m	01.02.2018–16.05.2018
W1 (Najdena jama)	Schlumberger, 100 m Hobo, conductivity	08.01.2015–16.05.2018 25.08.2016–16.05.2018
W2 (Gradišnica)	Hobo, pressure, 76 m Hobo, conductivity	08.01.2015–11.05.2018 25.08.2016–11.05.2018
W3 (Gašpinova jama)	Hobo, pressure, 76 m Hobo, conductivity Hobo, conductivity	28.01.2015–20.05.2018 28.01.2015–22.04.2015 25.08.2016–20.05.2018
E1 (Logarček)	Hobo, pressure, 76 m Hobo, conductivity	12.02.2015–11.05.2018 11.06.2015–07.06.2017
E2 (Vetrovna jama pri Laški kukavi)	Hobo, pressure, 76 m Hobo, conductivity	16.01.2015–20.05.2018 16.01.2015–20.05.2018
E3 (Brezno pod Lipovcem)	Schlumberger, 10 m	20.12.2014–26.06.2017
H1 (Veliko brezno v Grudnovi dolini)	Hobo, pressure, 76 m Hobo, conductivity	06.02.2015–16.05.2018 30.06.2016–16.05.2018
H2 (Andrejevo brezno 1)	Hobo, pressure, 30 m	30.12.2016–16.05.2018
S1 (springs Močilnik)	Hobo, conductivity	16.09.2016–09.05.2018
S2 (springs Retovje)	Hobo, conductivity	16.09.2016–13.03.2018
S3 (springs Bistra)	Hobo, conductivity	16.09.2016–09.05.2018

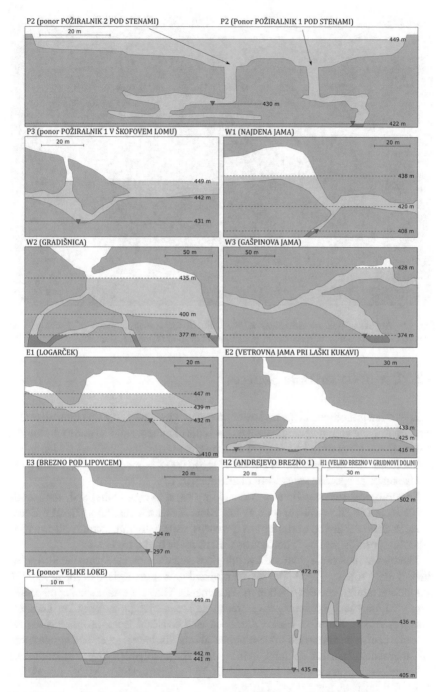

Fig. 4.4 Schematic view of observation points in selected ponors and caves. Red triangles indicate instrument position. Blue indicates water level, with dark blue representing the lowest and pale blue the highest measured water level. Dashed lines indicate the position of the lowest and highest water level and estimated levels of the overflow (obtained and adapted from [7])

Fig. 4.5 Acquiring field measurements of the Unica River: **a** measurements of discharge with ADCP (*Photo* F. Gabrovšek); **b** manual measurement of SEC (*Photo* M. Blatnik)

4.2 Use of Other Data Sources

Besides what is included in the review of past work (see Chap. 3), other data were also used: geological and speleological maps, and data of meteorological and hydrological observations. Most of the geological data originate from the 1960s to 1980s. In the 1960s, a basic geological map in a scale of 1:100,000 was made [17, 19], while more precise maps of specific areas were made later [9, 10].

Speleological investigations were made by local caving clubs; and, the results are collected in the cave register at the Karst Research Institute [7]. The documents include detailed descriptions of the caves; especially important is information regarding position and morphology, as well as maps of the caves.

The core of meteorological and hydrological data was obtained from the Slovenian Environment Agency [2, 3]. The meteorological data is comprised of the amount of continuous rainfall and distribution measurements. They are established at dispersed places within the Ljubljanica River recharge area: Ljubljana, Vrhnika, Logatec, Postojna, Juršče, Korošče, Bloke, Cerknica, and Sviščaki (Fig. 4.6). Hydrological measurements are comprised of continuous stream discharge, level, and temperature measurements of surface streams. In the area of the Ljubljanica River recharge area, these stations are on the rivers Ljubljanica, Ljubija, Bistra, Unica, Malni, Logaščica, Pivka, Cerkniščica, and Stržen (Fig. 4.6).

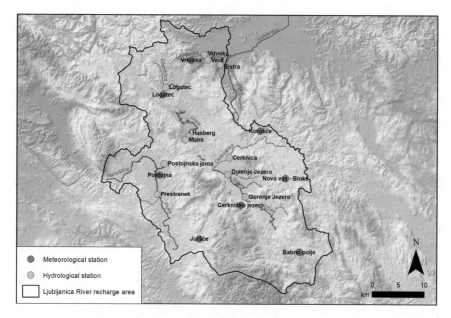

Fig. 4.6 The position of official meteorological and hydrological stations within the Ljubljanica River recharge area ([2, 3]; DEM data from [4])

4.3   Data Analysis and Interpretation

Several approaches were combined to obtain information on the geometry and functioning of the observed system based on the recorded data. Different plots were used to analyse the data. Typical events and responses at different positions were identified. Correlation plots were made to study the relations between the subsequent observation points. A careful qualitative inspection and heuristic interpretation of all hydrographs came out to be the crucial point in the workflow. Temperature and specific electrical conductivity measurements allowed estimations of transit times along the main flow pathways. The focus was also on the interaction between Planinsko Polje and the adjacent aquifer. Numerical models were used to verify the conceptual interpretation and to explore basic hydraulic relations in the epiphreatic zone. Information about the water level in different stages was interpolated and used to show the apparent groundwater table.

4.3.1   *Interpretation of Hydrographs*

In the present work, the time series presents hydrological and meteorological data obtained from the measurement network, the observation network set up in this study, and the network of the Slovenian Environment Agency (ARSOa, ARSOb). Records of the following parameters were used:

- relative water level (in meters above position of the instrument),
- absolute water level (in meters above sea level),
- water temperature (in °C),
- specific electrical conductivity of water (in μS/cm),
- rainfall (in mm/h or mm/day),
- stream discharge (in m³/s or l/s).

Water level hydrographs were the major source of information. Different plots of relative and absolute water level as a function of time were used to identify typical events and characteristic features on the water level hydrographs. Correlation plots between water levels at different positions were used to study the relations between the points. Inflection points on the water level hydrographs were often observed at a particular height at all events, when such a height was surpassed. These points, with a temporarily slower change of the water level, indicate the position of overflow channels, which were often, but not always, known from the cave survey [13] (Fig. 4.7). Through careful inspection of the water level hydrographs and correlation plots, and by taking into account the geometry of the known cave systems, the conceptual models of flood response and dominant flow pathways were built. Water level hydrographs were also used to obtain the spatial distribution and dynamics of the groundwater surface during various hydrological events. Of course, here one must keep in mind the limitations imposed by the limited number of observation points.

Water temperature hydrographs were analysed for annual and diurnal temperature cycles. Diurnal cycles imposed by the concentrated inflow of the Unica River

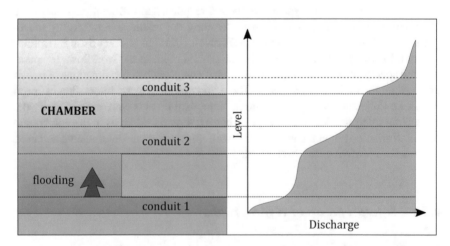

Fig. 4.7 A simple model of the flooding of a chamber that is draining through conduits at different levels and the resulting transitions between the concave and convex stage-discharge curve (adapted from [13])

allowed for estimation of transit times between consecutive points along the identified flow paths. These were estimated from the time lags of diurnal maxima and minima between the points. The diurnal temperature signal was also used as an identifier of overflow phenomena.

Specific electrical conductivity (SEC) hydrographs were also used as a natural tracer, where time lags between characteristic changes at different points were calculated. Furthermore, SEC hydrographs allowed basic for the estimation of different flow components at observation points and identification of possible pollution events.

The data on precipitation as a main driving force of all changes were plotted along the other hydrographs. The data from several stations of the national network [3] were used and combined. The response time and intensity was studied for different events. However, the main use of this data was to identify the duration and intensity of events.

The Unica River is the dominant inflow and direct driving force for the observed system. Its **stream discharge** and temperature hydrographs obtained from the national network [2] were related to all other observed hydrographs and used as inputs into numerical models.

Statistical analyses were also conducted on the observed parameters. They comprised descriptive statistics such as minimal, maximal, and mean values, calculations of standard deviations, moving averages, etc. Additional statistical analyses were made for the calculation of correlation of stage, temperature, and specific electrical conductivity records.

4.3.2 Conceptual and Numerical Hydraulic Models

One of the main goals of this work was to relate the geometry of the aquifer with the recorded hydrographs, and to infer its unknown geometry. To this extent, conceptual models were built and translated into corresponding numerical models, where at least a qualitative fit to the observed data was sought.

Models were built based on several assumptions:

- The flood response of the system is mainly controlled by a limited number of constrictions in the epiphreatic conduits.
- Turbulent flow in the channels and conduits is the dominant flow type in the epiphreatic zone.
- Transitions between open channel and full pipe flow (and vice versa) are common during flood events. Transitions are recorded as inflection points in the water level hydrographs [13] (Fig. 4.7).
- Large underground chambers may also play important role and cause inflections in the level curve.
- Epiphreatic flow is highly determined by the position of the overflow channels. Activation/deactivation of the overflow channels is also recorded as an inflection in the water level hydrograph.

The water level hydrographs were carefully inspected for inflection points. The known geometry of the caves was used to identify potential restrictions, overflow channels, and large storage voids (chambers) that could be related to inflections. Correlations between the different observation points, discharge of the Unica River, and precipitation records were used to identify common or related features on the water level hydrographs. In this way, a conceptual model of the conduit system was built and extended into the numerical model.

A numerical model was constructed within the EPA SWMM environment developed by the US Environmental Protection Agency [12]. SWMM was primarily developed to simulate urban sewage systems, but has found many applications to karst conduit networks [8, 13, 14, 18]. The model allows the simulation of many possible scenarios and permits the use of many elements of man-made and natural hydraulic systems. The model solves the full 1D Saint-Venant's equation [11] for a network of conduits positioned in space. To account for the transition from an open channel to a full pipe flow, a surcharge algorithm is used [20].

In this work, the full dynamical wave solution was used. The model only accounts for the flow in conduits, which limits its use to specific conditions of flood response in conduit dominated karst aquifers. Figure 4.8 shows the SWMM's Graphical User Interface, which is used to construct the models and view the results. A typical model in this work includes a system of links representing channels and conduits, and nodes and storage reservoirs representing junctions and chambers (Fig. 4.9). Only parts of the system with observation points were modelled. The aim was to use the minimal number of elements needed to capture the dynamics recorded by the

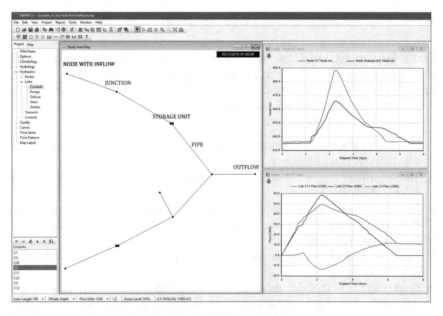

Fig. 4.8 The graphical user interface in EPA SWMM software

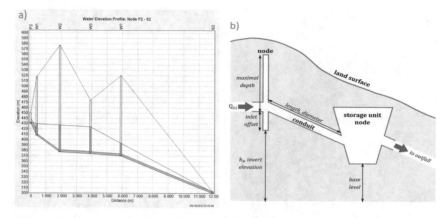

Fig. 4.9 Conceptual model constructed in SWMM: **a** longitudinal profile of conceptual model; **b** schematic view of elements and parameters used in conceptual model (adapted from [13])

observation network. The size of the constraining conduits as well as the position and size of epiphreatic overflows were varied to obtain a good and robust qualitative (in some cases also quantitative) match between the models and observations. Although several authors [8, 13, 14] have used inverse algorithms to optimise the model to the recorded field results, it was not done so in this work. The system is simply too undetermined to do so. A time series of the Unica River discharge was used as an input into most models. Results of the outflow measurements along the eastern border of the polje were used to get the actual inputs into the modelled sub-systems.

SWMM also proved to be a valuable learning tool. Apart from modelling the sub-systems in the observed area, SWMM was also used as a "sand-box", where various scenarios with combinations of voids, overflow channels and restrictions were modelled to learn about possible water level hydrographs obtained at various positions in the model.

4.3.3 Water Level Interpolation

The plotting of potentiometric surface in karst might be questionable, particularly when the observation points are scarce. Nevertheless, water level data from all observation points was used to discuss the piezometric surface in the area. For this, ArcGIS software was used and the Natural Neighbour interpolation was applied. The algorithm used by the Natural Neighbor interpolation tool finds the closest subset of input samples to a query point and applies weights to them based on proportionate areas to interpolate a value [22]. It is also known as Sibson or "area-stealing" interpolation [1]. The result is a contour map of the water table (Fig. 4.10a, b). Despite the above mentioned warning, the approach still enables simple interpretation of the water table and gradient during various hydrologic conditions.

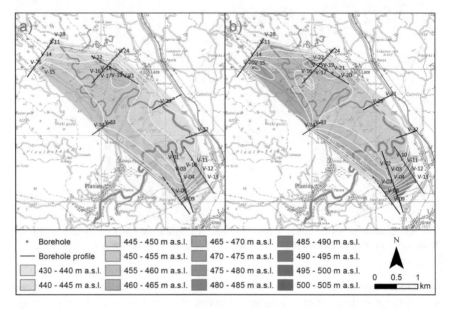

Fig. 4.10 Interpolations of piezometric measurements in low (**a**) and high (**b**) water level (obtained from [6])

A similar visualization was also made for measurements published by Breznik [6]. The measurements of the water level in various stages were made in boreholes. They were drilled in many places at the border of Planinsko Polje during a comprehensive study of the polje's floor structure and composition (Fig. 4.10).

References

1. ArcGIS for Desktop (2018) How Natural Neighbor works (online). Available from http://des ktop.arcgis.com/en/arcmap/10.3/tools/spatial-analyst-toolbox/how-natural-neighbor-works. htm. Accessed 7 Apr 2020
2. ARSO (2018a) Hydrological archive (online). Available from http://vode.arso.gov.si/hidarhiv/. Accessed 7 Apr 2020
3. ARSO (2018b) Meteorological archive (online). Available from http://meteo.arso.gov.si/met/ sl/archive/. Accessed 7 Apr 2020
4. ARSO (2018c) Lidar data fishnet (online). Available from http://gis.arso.gov.si/. Accessed 7 Apr 2020
5. Blatnik M, Frantar P, Kosec D, Gabrovšek F (2017) Measurements of the outflow along the eastern border of Planinsko Polje, Slovenia. Acta Carsologica 46/1:83–93. https://doi.org/10. 3986/ac.v46i1.4774
6. Breznik M (1961) Akumulacija na Cerkniškem in Planinskem polju. Geologija 7:119–149
7. Cave Register (2018) Cave Register of the Karst Research Institute ZRC SAZU and Speleological Association of Slovenia. Postojna, Ljubljana

8. Chen Z, Goldscheider N (2014) Modeling spatially and temporally varied hydraulic behavior of a folded karst system with dominant conduit drainage at catchment scale, Hochifen-Gottesacker, Alps. J Hydrol 514:41–52. https://doi.org/10.1016/j.jhydrol.2014.04.005
9. Čar J, Gospodarič R (1983) O geologiji krasa med Postojno, Planino in Cerknico. Acta Carsologica 13:91–106
10. Čar J (1981) Geološka zgradba požiralnega obrobja Planinskega polja. Acta Carsologica 10:75–106
11. Dingman SL (2015) Physical hydrology, 3rd edn. Waveland Press, Long Grove, p 643
12. EPA (2014) Storm Water Management Model (SWMM). US Environmental Protection Agency
13. Gabrovšek F, Peric B, Kaufmann G (2018) Hydraulics of epiphreatic flow of a karst aquifer. J Hydrol 560:56–74. https://doi.org/10.1016/j.jhydrol.2018.03.019
14. Kaufmann G, Gabrovšek F, Turk J (2016) Modelling flow of subterranean Pivka River in Postojnska Jama, Slovenia. Acta Carsologica 45/1:57–70. https://doi.org/10.3986/ac.v45i1.3059
15. ONSET Comp (2018a) HOBO® U20 Water Level Logger (U20-001-0x and U20-001-0x-Ti) Manual (online). Available from http://www.onsetcomp.com/files/manual_pdfs/12315-J%20U20%20Manual.pdf. Accessed 7 Apr 2020
16. ONSET Comp (2018b) HOBO® U24 Conductivity Logger (U24-001) Manual (online). Available from https://www.onsetcomp.com/files/manual_pdfs/15070-J%20U24-001%20Manual.pdf. Accessed 7 Apr 2020
17. Osnovna geološka karta 1:100.000, List Postojna (1963) Zvezni geološki zavod, Beograd
18. Peterson E, Wicks C (2006) Assessing the importance of conduit geometry and physical parameters in karst systems using the storm water management model (SWMM). J Hydrol 329:294–305. https://doi.org/10.1016/j.jhydrol.2006.02.017
19. Pleničar M (1970) Tolmač osnovne geološke karte 1:100.000, List Postojna. Zvezni geološki zavod, Beograd
20. Rossman LA (2015) Storm water management model user's manual version 5.1. U.S. Environmental Protection Agency, p 352
21. Schlumberger Water Services (2018) Diver Manual (online). Available from https://www.daiki.co.jp/Manual/Diver%20(GB)%20HL341v3.pdf. Accessed 7 Apr 2020
22. Sibson R (1981) A brief description of natural neighbour interpolation. In: Barnett V (ed) Interpreting multivariate data. Wiley, New York, pp 21–36

Chapter 5
Results and Discussion

Although the results of this work indicate good hydraulic connectivity throughout the observed system, it is still reasonable to divide it into several sub-systems, which can be treated separately. This is also supported by past research [10, 12, 18, 21] which identified regions with dominant flow connections (Fig. 5.1). Such divisions also makes further discussions easier. Therefore, three such subsystems are analysed separately in the following sections:

- Section 5.1: the subsystem which is mainly recharged over the eastern ponor zone of Planinsko Polje (ponor Velike loke, cave Logarček, cave Vetrovna jama pri Laški kukavi, and cave Brezno pod Lipovcem).
- Section 5.2: the subsystem which is primarily recharged through the northern ponor zone of Planinsko Polje (ponors Pod stenami, ponors Škofov lom, cave Najdena jama, cave Gradišnica, and cave Gašpinova jama).
- Section 5.3: the subsystem on the southern side of the Idrija Fault Zone, recharged from the Hrušica Plateau (caves Veliko brezno v Grudnovi dolini and Andrejevo brezno 1).

For each subsystem, a detailed description of water level dynamics and inferred flow directions, geometry of the epiphreatic flow system, and identification of possible flow barriers is presented. Assumptions are tested with a numerical hydraulic model (SWMM).

The three springs of the Ljubljanica River are analyzed in a separate section (Sect. 5.4). According to temperature and specific electrical conductivity measurements, interpretation of various high water events and possible sources of water is presented.

Analyses of the subsystems are followed by a description of the general characteristics of the water level (Sect. 5.5), temperature, and specific electrical conductivity (Sect. 5.6). The general pattern of water level increase and recession during high water events is also presented. This is followed by interpretation of the water temperature and specific electrical conductivity pattern during flood events. Both

M. Blatnik, *Groundwater Distribution in the Recharge Area of Ljubljanica Springs*, Springer Theses, https://doi.org/10.1007/978-3-030-48336-4_5

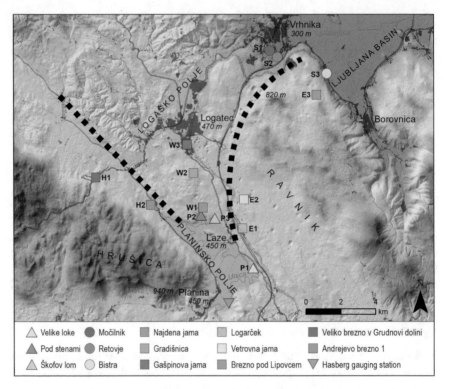

Fig. 5.1 Rough division of study area into subsystems according to previously recognized main water courses. Subsystems are delimited with black dashed lines (DEM data from ARSO [2]; Cave data from Cave Register [5])

water temperature records and specific electrical conductivity records are used for calculating transit times of water between selected consecutive observation points.

Based on all of the analyses, an improved interpretation of the groundwater directions and the influence of the geological structure is presented (Sect. 5.7). A map is included, which consists of previous and new findings.

In order to make the descriptions clearer, the names of the observation points have been given abbreviations (Table 5.1). "P#" represents ponors, "W#" represents caves related to the western regional groundwater flow, "E#" represents caves related to the eastern regional groundwater flow, "H#" represents caves recharging from the Hrušica Plateau, and "S#" represents the springs of the Ljubljanica River. Because they are completely connected hydraulically and behave similarly, ponors Požiralnik 1 pod stenami and Požiralnik 2 pod stenami share the same label (Table 5.1).

Table 5.1 List of studied ponors, caves, and springs with their abbreviations

Name of observation point	Abbreviation
Velike loke (ponor)	P1
Požiralnik 1 and Požiralnik 2 pod stenami (ponor)	P2
Požiralnik 1 v Škofovem lomu (ponor)	P3
Logarček	E1
Vetrovna jama pri Laški kukavi	E2
Brezno pod Lipovcem	E3
Najdena jama	W1
Gradišnica	W2
Gašpinova jama	W3
Veliko brezno v Grudnovi dolini	H1
Andrejevo brezno 1	H2
Veliki Močilnik (spring)	S1
Retovje—Izvir pod skalo (spring)	S2
Bistra—Galetov izvir (spring)	S3

5.1   System Related to the Eastern Ponor Zone of Planinsko Polje

The system draining Planinsko Polje can be, with some reservation, divided into the system related to the northern ponor zone, discussed in Sect. 5.2, and the system fed by the eastern ponors (Figs. 5.1 and 5.2). Although both systems seem to interact, they are, at this point, treated separately. The latter system includes caves E1 (Logarček) and E2 (Vetrovna jama pri Laški kukavi). The last cave is E3 (Brezno pod Lipovcem), close to the most eastern group of the Ljubljanica River springs S3 (spring of Bistra) (Fig. 5.2).

5.1.1   Hydraulic Relations Between Eastern Ponor Zone of Planinsko Polje (P1) and Caves Logarček (E1) and Vetrovna Jama Pri Laški Kukavi (E2)

At some observation points it is not possible to assure the permanent underwater position of the instrument without diving. Such is the case at E1 (Logarček) and the ponor P1 (Velike Loke) (Fig. 5.3). Particularly, in E1 (Logarček), the position of the instrument is about 20 m above the lowest water stand. This is at 410 m a.s.l., which is 6 m below the level of the observation point in E2 (Vetrovna jama pri Laški kukavi). Therefore, the interpretation of the hydraulic connection between E1 (Logarček) and E2 (Vetrovna jama pri Laški kukavi) is questionable during low water conditions.

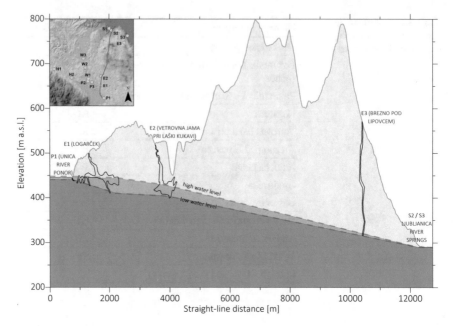

Fig. 5.2 Longitudinal profile of low and high water level between the eastern ponors of Planinsko Polje and Ljubljanica River springs

More reliable results are available for high water events, when water level is high enough to be recorded at all observation points. When Planinsko Polje is at least partially flooded, all changes in water level are transferred downstream to the locations of E1 (Logarček) and E2 (Vetrovna jama pri Laški kukavi) (Fig. 5.4). Measurements of temperature and SEC in both caves also show very similar results (Fig. 5.11).

The straight-line distance between E1 (Logarček) and the nearest ponor zone on the polje (ponor Dolenje loke) is about 1500 m (Fig. 5.3). The head difference between Planinsko Polje (P1—Velike loke) and E1 (Logarček) decreases with increasing recharge; during high water events it drops to less than 3 m (Fig. 5.4). This makes the gradient less than 0.002. Such a small gradient presents almost flat surface of the water table and indicates that transmissivity in area between the eastern ponors of the Planinsko Polje and E1 (Logarček) is significantly higher than in the area downstream of E1 (Logarček).

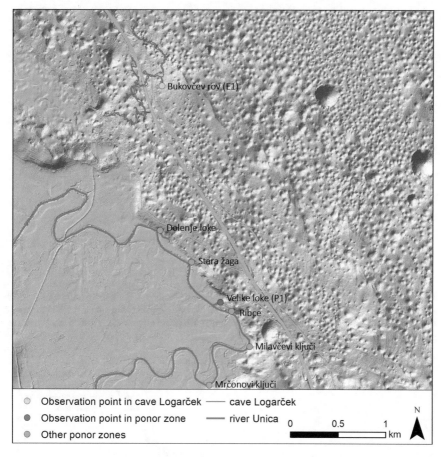

Fig. 5.3 Locations of observed (P1) and other ponor zones on eastern border of Planinsko Polje and cave Logarček with observation point (E1) in passage Bukovčev rov (DEM data from ARSO [2]; Cave data from Cave Register [5])

5.1.2 Water Level Relations Between Logarček (E1) and Vetrovna Jama Pri Laški Kukavi (E2)

During the initial stage of the flood event, no clear statement can be made about the hydraulic connection between both caves. The initial rise of level in E2 (Vetrovna jama pri Laški kukavi) (Fig. 5.5) is probably strongly related to the breakdown that controls the flow from the observation point towards the large chamber and channel leading to the outflow sump (Fig. 5.6). In E1 (Logarček), however, the water first needs to reach the instrument 22 m above the lowest stand (from 410 to 432 m a.s.l.). It continues to rise rapidly for another 7 m (Fig. 5.5) to the level of a large overflow channel (passage Skalni rov, 439 m a.s.l.) (Fig. 5.6). This makes a major inflection in the rising limb and allows for fast discharge of all excess water (Fig. 5.5). From there,

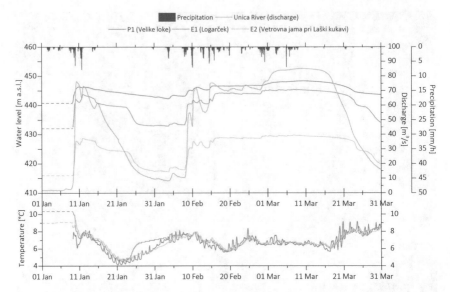

Fig. 5.4 Water level and temperature hydrograph during the high water event from beginning of 2016 in ponor P1 (Velike loke) and caves E1 (Logarček) and E2 (Vetrovna jama pri Laški kukavi). For comparison, discharge measurements of the Unica River are also shown. Dashed lines indicate periods when the instrument was out of water

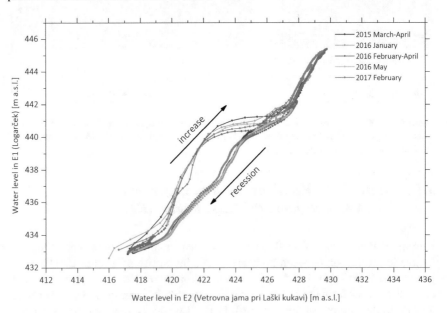

Fig. 5.5 The correlation between water levels at observation points Logarček (E1) and Vetrovna jama pri Laški kukavi (E2)

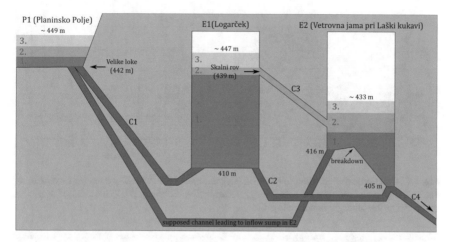

Fig. 5.6 Conceptual model of the assumed distribution of passages (C1, C2, C3, and C4) that connect P1 (Planinsko Polje) with caves E1 (Logarček) and E2 (Vetrovna jama pri Laški kukavi). The higher positioned inflow sump in E2 is most likely discharged directly from Planinsko Polje, whereas the lower could be connected to E1. The presented conditions were used for constructing the SWMM model. Numbers represent phases visible on water level hydrographs on Figs. 5.7a, 5.8a and 5.9a

all graphs show a close resemblance between points E1 and E2, indicating a good hydraulic connection (Fig. 5.8). It seems that the level in both caves is controlled by the transmissivity of the system beyond E2 (Vetrovna jama pri Laški kukavi), which also controls a relatively simultaneous recession of the level in both caves (Figs. 5.8 and 5.9).

For a better understanding of the system, a simple SWMM model was made. It resembles the major characteristics of the observed water level hydrographs. It is composed of nodes P1, E1, E2 (representing Planinsko Polje, caves Logarček and Vetrovna jama pri Laški kukavi, respectively) and an outlet connected with circular conduits (C1, C2, C3, C4) (Fig. 5.6). Nodes can be simple junctions or storage units. The water enters at the highest node (P1), representing a ponor. This has two outflows, one (C1) at the base flow, connected to the second node (E1), and one several meters above the junction divert, representing the polje's overflow. The second junction has a base flow channel (C2) towards E2, and an overflow channel (C3) connecting E1 and E2. E2 is drained with a single base channel (E4) (Fig. 5.6). Detailed descriptions of all elements in the model are presented in Appendix A. Figures 5.7b, 5.8b, and 5.9b show the results of the model.

Figure 5.7 shows the water level hydrographs at all three points. Figure 5.7a shows the actual high water event, whereas Fig. 5.7b shows a modelled high water event. The system is being filled until the head rises to the position of the polje and the excess of water is drained out by C1. The system remains full until the recharge drops to the value when the heads in the system cannot be sustained and the recession starts.

The correlation between heads at both internal junctions, which resemble one between E1 (representing cave Logarček) and E2 (representing cave Vetrovna jama

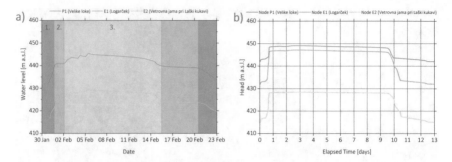

Fig. 5.7 Relation between water level in E1 (Logarček) and E2 (Vetrovna jama pri Laški kukavi) during a high water event: **a** the actual measurements from February 2017; **b** modelled dynamics, according to the settings shown on Fig. 5.6

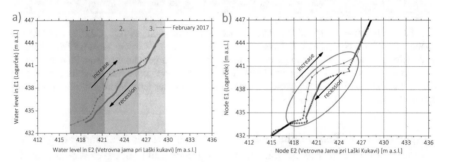

Fig. 5.8 The correlation between heads at observation points in E1 (Logarček) and E2 (Vetrovna jama pri Laški kukavi) during a high water event: **a** the actual measurements from February 2017; **b** results of modelled high water event, according to the settings shown on Fig. 5.6

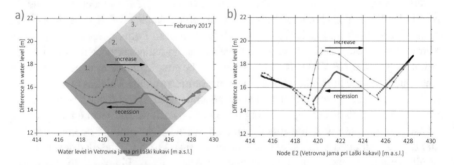

Fig. 5.9 Difference in heads in caves E1 (Logarček) and E2 (Vetrovna jama pri Laški kukavi) during selected high water event, according to head in E2 (Vetrovna jama pri Laški kukavi): **a** the actual measurements from February 2017; **b** results of modelled high water event, according to settings shown on Fig. 5.6

pri Laški kukavi) is shown on Fig. 5.8. Figure 5.8a shows the correlation during an actual high water event, whereas Fig. 5.8b shows the correlation of a modelled high water event. Both plots show a characteristic hysteresis caused by the overflow channel.

1. When E1 and E2 are connected only by C2, the head at both nodes increases in correspondence to the increasing recharge.
2. When water at E1 reaches the position of the overflow (C3) the increasing recharge flows as an open channel flow through C3 and fills the junction E2. For this reason, the head difference between E1 and E2 drops until the level at E2 reaches the position of an overflow.
3. From then on, both heads rise until the maximum.

The recession follows the rising curve until the position of the overflow, but the relation between the head at E1 and E2 in the reverse situation is different; the head drop at E2 is not so abrupt, which leads to the offset between the rising and recession part (Fig. 5.8).

Similar curves are observed for large events between E1 (Logarček) and E2 (Vetrovna jama pri Laški kukavi). The inflection is at the position of the overflow channel in E1 (Logarček). Even though there are other possible models which could explain the results, the simplest given above (Fig. 5.6) seems most plausible.

An important element of the correlation given in Fig. 5.8 is the hysteresis (marked in Fig. 5.8b), indicating a difference between the rising and recession limbs at both points. By testing the model, it is seen that the form of the hysteresis loop depends on the length/size of C3, which is related to storage in C3.

Further underground flow between E2 (Vetrovna jama pri Laški kukavi) and the springs of the Ljubljanica River is not well known. There is a cave, E3 (Brezno pod Lipovcem), which is partially flooded during the most intensive rainfall events. More about the dynamics in this cave is contained in Sect. 5.1.4. The drop of the water level between E2 (Vetrovna jama pri Laški kukavi) and the springs of the Ljubljanica River in various hydrologic conditions is between 105 and 129 m at the straight-line distance of 8100 m. This presents a gradient between 0.013 and 0.016, which is more than in the upstream groundwater flow (0.009–0.012).

5.1.3 Water Temperature and SEC Relations Between Logarček (E1) and Vetrovna Jama Pri Laški Kukavi (E2)

During low water conditions, both instruments in E1 (Logarček) and E2 (Vetrovna jama pri Laški kukavi) are out of water, so temperature and SEC records are irrelevant. When the instruments are in the water, two situations are possible (Fig. 5.10):

1. When there is no flow from Bukovčev rov to Skalni rov in E1 (Logarček) (Fig. 5.11), the instrument is in standing water. The temperature and SEC records

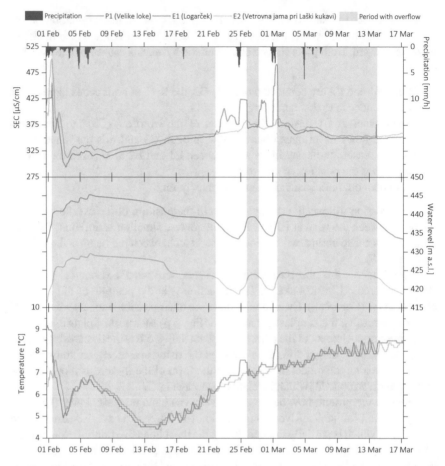

Fig. 5.10 Water level, temperature, and specific electrical conductivity hydrograph shows a good match between E1 (Logarček) and E2 (Vetrovna jama pri Laški kukavi) when water level exceeds 439 m a.s.l.—the level at which water from passage Bukovčev rov overflows into main passage Skalni rov. Periods with active flow through passage Bukovčev rov are marked with blue belts. Presented high water event occurred at the beginning of 2017

are very distinct from that obtained at E2 (Vetrovna jama pri Laški kukavi) (Fig. 5.10). They probably represent the signal of local infiltration. Water temperature is slowly adapting to cave air temperature, while SEC records have periodic increases (Fig. 5.10). In winter, manual measurements of conductivity in some puddles reached 2000 μS/cm, probably caused by salt water infiltration from the nearby motorway. In E2 (Vetrovna jama pri Laški kukavi), during low water conditions, a stream of water is active. There, diurnal temperature oscillations and relatively stable SEC values were recorded (Fig. 5.10).

2. When the level in E1 (Logarček) reaches 439 m a.s.l., the overflow from Bukovčev rov to Skalni rov becomes active (Fig. 5.11), and simultaneous changes of water

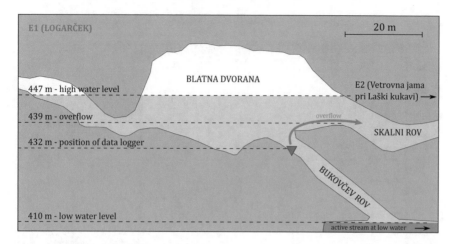

Fig. 5.11 Cross section of E1 (Logarček) in the region of the instrument (marked by the red triangle). The lowest position of water is at 410 m a.s.l. (dark blue), and during low water conditions, the active stream at the floor of passage Bukovčev rov is present. During a flood event, water level increases to that extent, then overflows into the main passage, Skalni rov (overflow marked with blue arrow), and flows towards E2 (Vetrovna jama pri Laški kukavi) (Cave data from Cave Register [5])

temperature and SEC at both observation points are recorded. After initial sudden change of temperature, diurnal oscillation in both caves occur (Fig. 5.10). In comparison to temperatures on the polje, they are slightly damped because of heat exchange with rock and air. SEC measurements first record a sudden increase (Fig. 5.10), when saturated stored water is flushed out. After that, SEC is controlled by dilution through the rainfall water and also by the mixing of fast flow water (allogenic recharge and water from polje, which is less saturated with respect to carbonates) and more saturated slow flow water (mainly epikarstic and vadose) components.

5.1.4   Dynamics of Water Level and Temperature in Cave Brezno Pod Lipovcem (E3)

Cave E3 (Brezno pod Lipovcem) is the northernmost cave with an access to the groundwater in the aquifer between Planinsko Polje and the springs of the Ljubljanica River. It is an approximately 280 m deep, dominantly vertical, cave located about 1 km southwest from springs S3 (Bistra) and 2.5 km southeast from the group of springs S2 (Retovje) (Fig. 5.14). Most of the year, the deepest known point of the cave is dry. But, during high water events, it is flooded and during that time some mud is deposited.

Measurements of water temperature and water level were established from January 2015 to June 2017 (Fig. 5.12). The instrument was installed at the deepest point of

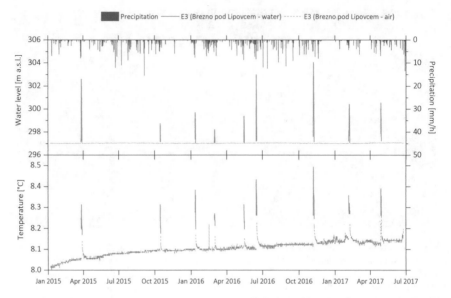

Fig. 5.12 Time series of water level and temperature measurements in E3 (Brezno pod Lipovcem). Most of the time the instrument was above the water level, which is marked with dashed lines

the cave at an elevation of 297 m a.s.l. In that period, 10 events were recorded when groundwater increased to the level of the instrument. Increase of the water level varied from 0.1 to 7 m above the position of the instrument (Figs. 5.12 and 5.13). The duration of each high water event was short, usually between 25 and 70 h (Fig. 5.13). The number of high water events in E3 (Brezno pod Lipovcem) was much smaller than in other studied caves, and their duration was much shorter as well.

Because of the observation point was positioned above the level of the low-flow water table, the hydrograph of high water level events is not complete (Figs. 5.13 and 5.15). It shows only the flood peaks with duration between 25 and 55 h. Water level hydrographs in E3 (Brezno pod Lipovcem) show similar dynamics to hydrographs of the spring discharge of the Ljubljanica River. However, peaks of the water level are always less than 2 h delayed with respect to the peak discharge of the Ljubljanica River (Figs. 5.13 and 5.15).

The flood levels in E3 (Brezno pod Lipovcem) probably show the rise of water level in the area of the springs. Such a rise also results in the flooding of collapse dolines near the town of Vrhnika (Fig. 5.14), which matches the highest water level in E3 (Brezno pod Lipovcem). Similar dynamics are also significant for the activity of the highest located overflow springs S2 in the group Retovje (Malo Okence and Veliko Okence springs) (Fig. 5.14).

Temperature records mostly show stable air temperatures between 8 and 8.14 °C (Fig. 5.12). Interestingly, it is about 2 °C lower than the average temperature in the neighbouring town of Vrhnika. The cave is poorly ventilated with many narrow

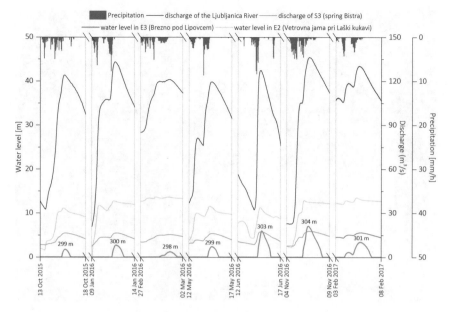

Fig. 5.13 Water level and spring discharge hydrograph of seven selected high water events in E3 (Brezno pod Lipovcem) compared to the water level in the upstream location of E2 (Vetrovna jama pri Laški kukavi), discharge of the springs S3 (Bistra) located downstream, and total discharge of the Ljubljanica River. Numbers indicate maximum water level in E3 (Brezno pod Lipovcem) during each high water event

channels leading to a large chamber at the lowest known part of the cave. As such, it cannot contain a cold trap. The reason for the low temperature is that the infiltration to the system is at an elevation above 550 m a.s.l. Because decrease of air temperature with altitude (–6.6 °C/km) is much higher than increase of temperature of percolating water (approximately 2.5 °C/km) [3], the rock at the base of the massifs with high elevation is typically lower than the outside temperature at the same elevation.

During flood events, the water temperature is a bit higher than the local air temperature, which is well below temperatures in the spring. This could indicate that the water that reaches the position of the instrument represents local water, which is only partially mixed with event waters from the poljes (Fig. 5.15).

5.2   System Related to the Northern Ponor Zone of Planinsko Polje

Past studies (Gospodarič and Habič 1976) [4, 10, 20] indicate that water from the ponors at the northern border of Planinsko Polje (P2 and P3) flows through the system of caves W1 (Najdena jama), W2 (Gradišnica), and W3 (Gašpinova jama) towards Ljubljanica River springs S1 (Močilnik) and S2 (Retovje) (Fig. 5.16). Results obtained in this part of the system are discussed in this section.

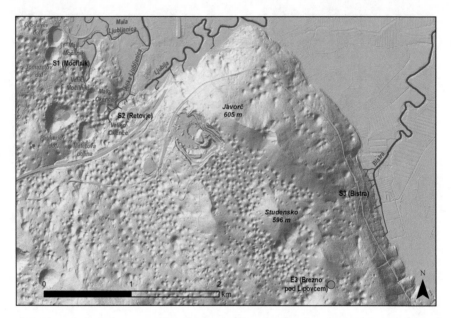

Fig. 5.14 Locations of cave E3 (Brezno pod Lipovcem), periodically flooded collapse dolines near Vrhnika (Meletova dolina, Paukarjev dol and Grogarjev dol), and springs of the Ljubljanica River (DEM data from ARSO [2])

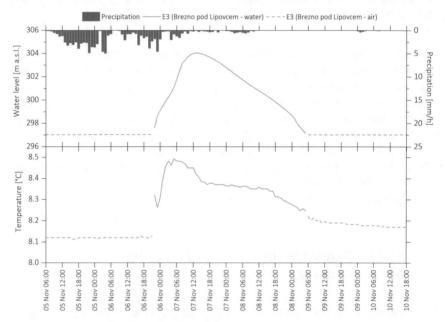

Fig. 5.15 Dynamics of selected high water event from November 2016 in cave E3 (Brezno pod Lipovcem), caused by an intensive rain event

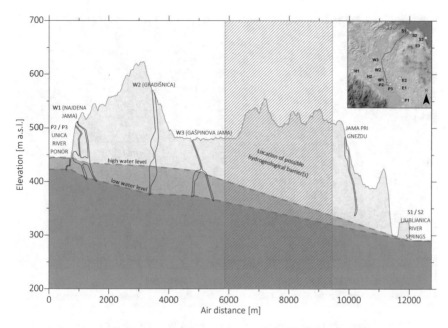

Fig. 5.16 Longitudinal profile of low and high water level between northern ponors of Planinsko Polje and Ljubljanica River springs. Red stripes denote location of possible geological barriers

5.2.1 Hydraulic Relations Between Northern Ponor Zone of Planinsko Polje (P2 and P3) and Najdena Jama (W1)

The northern part of Planinsko Polje has a large number of ponors, which are active when the discharge of the Unica River surpasses the outflow capacity of the eastern ponor zone. In this study, water level observations in three such ponors were established. The longest observation period (from January 2015 to November 2017) was in a ponor cave, Požiralnik 2 pod stenami, where water level and temperature were recorded. The observation point was positioned at 430 m a.s.l., reached by water only during flood events. For this reason, two additional observation points in neighbouring ponors were established, but for a shorter observation period. One of them (Požiralnik 1 pod stenami) is located 30 m from the previous. Both locations are presumably hydraulically connected, but the observation point in the latter is lower (422 m a.s.l) and always below the water. Because the point is permanently under water, SEC was also recorded.

Both ponors belong to the ponor zone "Pod stenami" (P2) (Fig. 5.17). The second additional point is part of the ponor zone "Škofov lom" (P3) (Fig. 5.17), about 700 m east from previous two. It is activated only at sufficiently high discharge, when water in the sinking channel rises above the level of a higher positioned channel leading to Škofov lom (Fig. 5.17). There, the observation point is positioned in the ponor cave Požiralnik 1 v Škofovem lomu, which ends with water at an elevation of 433 m a.s.l.

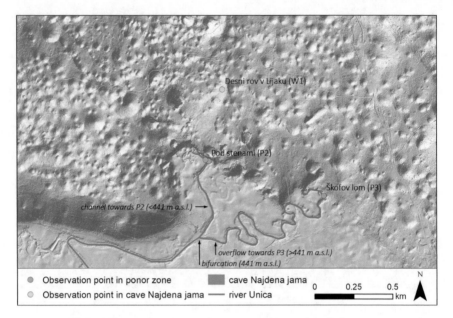

Fig. 5.17 Locations of the northern ponors of Planinsko Polje (P2 and P3) and previously or recently positioned observation points in W1 (Najdena jama) (DEM data from ARSO [2]; Cave data from Cave Register [5])

There are many passages with access to the groundwater in W1 (Najdena jama). Their position during low water level varies from 397 to 417 m a.s.l [17]. An observation point was positioned in the passage "Desni rov v Lijaku" at 408 m a.s.l. (Fig. 5.17). Except for during the longest dry period, the instruments were below the water level. The records were compared to records from 2006 and 2007 that were obtained in another sump, "Vipero nero" [21]. The sump is approximately 400 m northeast from Desni rov v Lijaku and about 8 m lower (Fig. 5.17).

The straight-line distance between ponor zone P2 (Pod stenami) and observation point Desni rov za Lijakom in Najdena jama (W1) is about 450 m. During low water, the head drop between both points is about 14 m, and 12 m during high water. This presents a gradient of 0.027. The onset of each large event shows a rapid increase of the water level in the ponor zone P2 (Pod stenami) (Fig. 5.18). However, the steep rise in the ponor precedes the rise in W1 (Najdena jama) by 4–8 h. Figure 5.18a, b show that the delay between both locations is also related to the intensity of the event. As expected, the faster the rise at both stations, the shorter the delay, as also seen in Fig. 5.19, which shows two events of different intensity.

Considering that the vigorous response at both locations is caused by the arrival of rapid floodwater, it is unlikely that the transit time from P2 (Pod stenami) to W1 (Najdena jama) would be so long. A much more plausible explanation is that W1 (Najdena jama) is dominantly fed by ponors P3 (Škofov lom), which responds later when the water level is higher in Planinsko Polje.

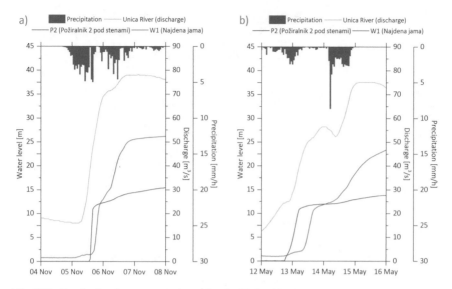

Fig. 5.18 Graphs showing response of water level at P2 (Požiralnik 2 pod stenami) and W1 (Najdena jama) and discharge of the Unica River during different seasons of the year: **a** quick response of water level during intensive rainfall in autumn 2016; **b** slower response during less intensive event in spring 2016

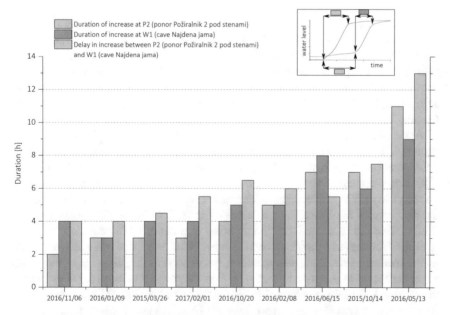

Fig. 5.19 Different duration of responses and delays between ponors P2 (Pod stenami) and W1 (Najdena jama) during various high water events. Durations are calculated for period of increase from base water level to first inflection point, shown in inset

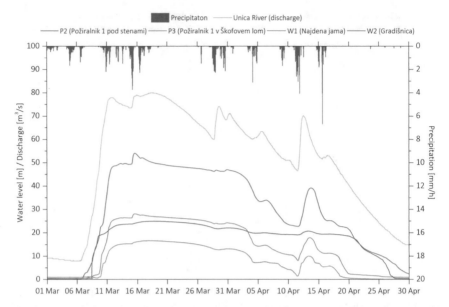

Fig. 5.20 High water event from spring 2018 with simultaneous dynamics in P3 (Škofov lom) and W1 (Najdena jama)

A high water event from March to April 2018 was monitored in both ponor zones on the northern border of Planinsko Polje (Fig. 5.20). As observed before, ponor zone P2 (Pod stenami) preceded W1 (Najdena jama). With several hours of delay, overflow towards ponor zone P3 (Škofov lom) occurred, which resulted in immediate response in W1 (Najdena jama) (Fig. 5.50). Moreover, the water level hydrographs at W1 (Najdena jama) and P3 (Škofov lom) strongly correlate during activity of P3 (Škofov lom) (Fig. 5.20).

At the end of February and beginning of March 2017, several moderate rain events occurred. Three of them are marked with arrows on Fig. 5.21. Following the rain pulse on February 24th, the discharge of the Unica River increased to about 35 m³/s. This caused a moderate response of the water level in P2 (Pod stenami), but almost no response in W1 (Najdena jama). The following rain pulse on March 1st increased the discharge of the Unica River to about 47 m³/s. This flooded the ponor zone P2 (Pod stenami) and activated a not yet observed ponor zone P3 (Škofov lom), which is recharged through a higher positioned channel in the polje. The water level response in W1 (Najdena jama) was also recorded, but as soon as the discharge of the Unica River decreased to about 43 m³/s, the level in the cave also dropped. Following the third pulse of the rain on March 4th, the discharge of the Unica River rose to about 57 m³/s, which caused a high response in W1 (Najdena jama), where water reached the level of the main system of galleries. The presented series of three rain events (Fig. 5.21) nicely demonstrates the activation of the northern ponors and their relation to W1 (Najdena jama). It suggests a poor connection between the ponor zone P2 (Pod

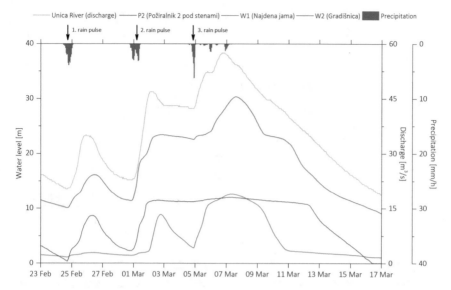

Fig. 5.21 Graph showing the beginning of 2017, when a series of rain events of various intensities caused different dynamic of water in selected ponors and caves

Stenami) and W1 (Najdena jama) and proves a dominant recharge of Najdena jama through ponor P3 (Škofov lom).

5.2.2   Dynamics of Water Level in Najdena Jama (W1)

Measurements at the observation point Vipero nero (Fig. 5.17) from 2006 to 2007 and recent measurements in the passage Desni rov v Lijaku show that water level responds at each moderate or high water event. A typical, small, initial rise of the level (usually less than 2 m) is followed by a high response when the discharge of the Unica River surpasses 45 m^3/s and activates the ponor zone P3 (Škofov Lom) (Fig. 5.20). The water level at both points immediately rises to the level of the main system of spacious passages (Fig. 5.24). In this situation, the water level in the whole cave is uniform.

Figure 5.22 shows two similar high water events. Figure 5.22a shows a high water event from May 2006 and was measured at observation point Vipero nero. Figure 5.22b shows a high water event from March 2015 that was measured at observation point Desni rov v Lijaku. Both plots show a small initial response of the water level. This change of water level is from 1 to 2 m (blue circles on Fig. 5.22a, b), so the gradient between both observation points is maintained. The main increase in the water level occurs later, when ponor zone P3 (Škofov lom) activates. The rising and recession limbs of the water level hydrographs show inflection at the level of large galleries, which effectively convey all the excess of flow (grey lines on Fig. 5.22a,

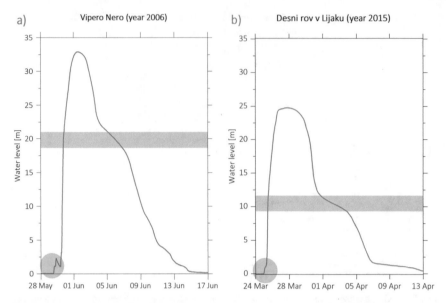

Fig. 5.22 Comparison of two similar high water events registered at observation points in sumps Vipero nero (year 2006) and Desni rov v Lijaku (year 2015). Blue circle denote initial water level response. Grey lines denote the position of the inflection point at the position of the overflow through the most transmissive zone. Note that the observation point in Vipero Nero was located approximately 8 m below the point in Desni rov v Lijaku

b). The relative water level in both plots differs for about 8 m, which presents a relative difference in the altitude of the observation points. The shape of the water level hydrograph differs slightly due to different characteristics of each flood event.

Figure 5.23a shows the rates of increase and decrease of the water level in Najdena jama during three selected flood events. It shows typical behaviour resulting from a steep rising limb to a gently sloping limb. At elevations around 420 m a.s.l., there is a clear minimum of the rate of water level increase and decrease. This is the elevation of the inflection points in Fig. 5.22 and indicates the level of highly transmissive large conduits (marked with red arrow on Fig. 5.23). Figure 5.23b shows the point cloud of rates of increase and decrease of the water level for the complete data set. The rates were calculated as the ratio between the change of water level and the time interval between two consecutive measurements. For a clearer view, a moving average taken along 51 points is shown by the red line on Fig. 5.23b. It shows a clear minimum at the same position (420 m a.s.l.). Because many events of all possible durations and magnitudes are included, it may be concluded that the minimum reflects the local geometry (Fig. 5.24).

Some questions remain about the water levels, recharge, and internal dynamics of flow in Najdena jama. The cave has been accurately surveyed and the vertical position of sumps should, in principle, be correct. On the other hand, the cave is too

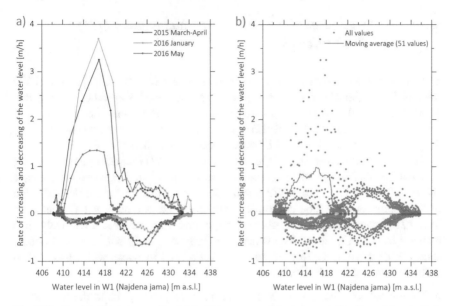

Fig. 5.23 Graphs showing rate of water level increase and decrease as a function of the absolute water level: **a** three selected flood events; **b** all flood events combined

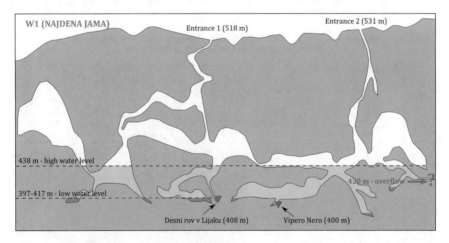

Fig. 5.24 Longitudinal profile of W1 (Najdena jama) with marked positions of observation points, low and high water level, and elevation with overflow (Cave data from Cave Register [5])

complex to capture the flood dynamics with a single measurement site. For this, at least 5–6 observation points are needed to get definitive results.

5.2.3 Hydraulic Relations Between Najdena Jama (W1) and Gradišnica (W2)

The straight-line distance between observation points W1 (Najdena jama) and W2 (Gradišnica) is about 2000 m. During low water, head drop between both observation points is about 30 m (Fig. 5.25). During each high water event, the water level in the lower positioned W2 (Gradišnica) rises for up to almost 60 m and decreases the drop of the water level from W1 (Najdena jama) to less than 5 m (Fig. 5.25).

Section 5.2.1 describes the delayed response in cave W1 (Najdena jama) in respect to ponor P2 (Pod stenami). The response is also delayed in comparison to the cave W2 (Gradišnica). Even more, W2 (Gradišnica) responds even before the activation of the ponor zone P2 (Pod stenami). Figure 5.25 shows a high water event from May 2016, when a moderate event and evapotranspiration caused a relatively slow increase in the water level. The water level hydrograph of both caves and ponors show a difference in the time response of water level increase, which are marked with different colours (Fig. 5.25):

- Yellow colour: The level of W2 (Gradišnica) responds prior to the level in P2 (Pod stenami). Water level rises for about 10–15 m.
- Orange colour: When ponor zone P2 (Pod stenami) activates, the water level immediately rises for about 10 m. The water level in cave W2 (Gradišnica)

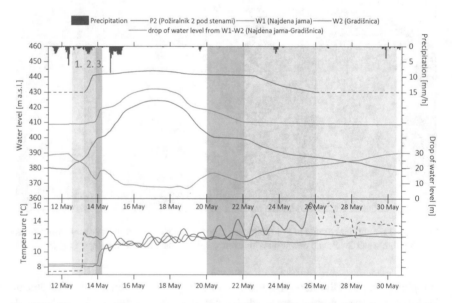

Fig. 5.25 Water level and temperature dynamics on Planinsko Polje and two following caves during a high water event in May 2016. The yellow belt denotes response in W2 (Gradišnica); the orange belt denotes response in ponors P2 (Pod stenami); the red belt denotes response in W1 (Najdena jama)

continues to rise to the position of the overflow through the passage Šerkov rov at 400 m a.s.l.

- Red colour: When the ponor zone P3 (Škofov lom) is activated, the level in W1 (Najdena jama) responds with a fast increase followed by an additional increase in W2 (Gradišnica).

A reasonable explanation for these dynamics can be given as the following (Fig. 5.26):

1. The initial increase of the water level in W2 (Gradišnica) is not caused by inflow through the northern ponors of Planinsko Polje (P2 or P3). The probable source of water is from the Hrušica Plateau region to the south of the Idrija Fault Zone, which is also supported by records in H1 (Veliko brezno v Grudnovi dolini) (Fig. 5.26 and Sect. 5.5). There, the response to the rain event occurs several hours before. Also possible is a hydraulic connection to E1 (Logarček) and E2 (Vetrovna jama pri Laški kukavi) fed through the eastern ponors of Planinsko Polje (P1).
2. When ponor zone P2 (Pod stenami) is activated, the outflow mainly bypasses W1 (Najdena jama), which shows no or a very small response, while in W2 (Gradišnica) the rate of the level rise increases. The level in W2 (Gradišnica) reaches the overflow position through passage Šerkov rov (Figs. 5.26 and 5.29). The overflow flattens the increase of the water level in W2 (Gradišnica).
3. As the water in the polje continues to rise, successive water is diverted along higher positioned channels towards ponor zone P3 (Škofov lom). After activation of this group of ponors, the level in W1 (Najdena jama) rises to the level of the

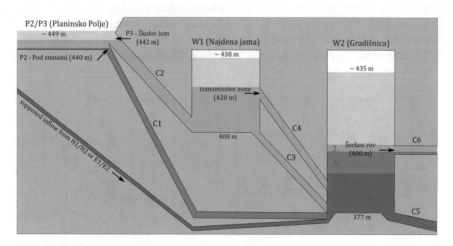

Fig. 5.26 Conceptual model of the assumed distribution of passages connecting Planinsko Polje (P2 and P3) with W1 (Najdena jama) and W2 (Gradišnica). W2 (Gradišnica) is discharged directly through lower positioned ponors (P2) and therefore active for a longer period. Presented conditions were used in the construction of the SWMM model. Numbers represent phases visible on water level hydrographs on Figs. 5.25 and 5.27a

most transmissive conduits (Figs. 5.24 and 5.26). It is not clear how much of the recharge from P3 (Škofov lom) flows through W1 (Najdena jama) and how much flows along the, yet unknown, conduits in its vicinity. However, the flow through the highly transmissive zone results in additional level rise in W2 (Gradišnica).

The recession is first marked by deactivation of ponor zone P3 (Škofov lom), which results in a drop in W1 (Najdena jama) to an almost initial level and to the overflow level in W2 (Gradišnica). Further decrease of inflow then results in deactivation of the ponor zone P2 (Pod stenami) and, therefore, in a gradual drop of the water level in W2 (Gradišnica).

The described concept was tested with a simplified SWMM hydraulic model. Nodes (caves) and connecting channels (cave passages) were distributed as they are presented on Fig. 5.26. Lower inflow from polje (P2) is directly connected to node W2 (representing Gradišnica), whereas higher inflow (P3) is connected with node W1 (representing Najdena jama). This node is further connected to W2 by a conduit C3 and an overflow channel C4 at a position 10 m higher. The polje gets an inflow similar to during the actual event from May 2016. A list of relevant parameters and their values is given in Appendix B.

As a result, the dynamics of water level shown on Fig. 5.27b agrees with the actual measurements shown on Fig. 5.27a. The head drop from W1 (Najdena jama) to W2 (Gradišnica) decreases dramatically during high water events. During the first phase of the high water event, the water level does not respond in W1 (Najdena jama), while the water level in cave W2 (Gradišnica) increases rapidly (yellow and orange belts on Fig. 5.27a). During the second phase, when the water level in W2 (Gradišnica) reaches a level of 400 m a.s.l., the drop of the water level increases for several meters (Fig. 5.27a). When ponor zone P3 (Škofov lom) is activated, the water level in W1 (Najdena jama) rises to the position of highly transmissive conduits, which increases the difference between both (red belt on Fig. 5.27a, b). Additional recharge to the W2 (Gradišnica) region also raises the level there, which results in a further decrease

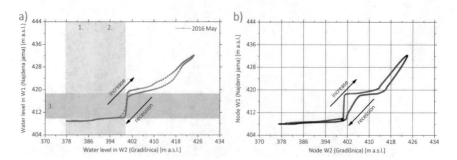

Fig. 5.27 Graphs showing relation between water level in caves W1 (Najdena jama) and W2 (Gradišnica) during a high water event in May 2016 (shown on Fig. 5.25): **a** actual measurements with yellow belt denoting response in W2 (Gradišnica), orange belt denoting response in ponors P2 (Pod stenami), red belt denoting response in W1 (Najdena jama); **b** modelled dynamics, according to settings shown on Fig. 5.26

of the head difference. At maximum, the head difference between both is only a few meters (Fig. 5.27a, b). Due to the reverse deactivation of the ponor zones, recession shows reverse but slower dynamics.

Water temperature shows a different pattern before the high water event. This indicates different characteristics of each observation point (Fig. 5.25). When a high water event begins, floodwater enters the system and mixes the water. When the water level reaches its highest extent and Planinsko Polje is flooded, diurnal temperature oscillations reach both stations (Fig. 5.25). In W1 (Najdena jama), diurnal temperature oscillations last for a shorter time and terminate along with an early drop in the water level (Fig. 5.25). Diurnal oscillations are present when there is an active flow of event water from the polje at the observation point. The time window of such oscillations may also vary from location to location within the same cave.

5.2.4 Hydraulic Relations Between Gradišnica (W2) and Gašpinova Jama (W3)

Although the geometry of the known parts of caves W2 (Gradišnica) and W3 (Gašpinova jama) is different, the hydrologic characteristics of them are very similar. A good hydraulic connection of both caves is indicated in a similar shape of the water level hydrograph (similar range and speed of water level fluctuation), while water temperature and SEC fluctuations are more similar only during high water events (Fig. 5.28).

Close inspection of the water level in W2 (Gradišnica) and W3 (Gašpinova jama) shows interesting inflections on the water level hydrograph during the rising and recession limbs (Fig. 5.28). This occurs when the level in W2 (Gradišnica) reaches the overflow level towards the passage Šerkov rov (Fig. 5.29) and temporarily slower increase and decrease of the water level at about 400 m a.s.l. is registered (Fig. 5.30). This is shown by an arrow on Fig. 5.29 and occurs during all events, where the water level rises to and beyond the overflow position.

To elucidate this phenomenon, a simple SWMM model has been constructed. The conceptual model is shown in Fig. 5.31. It consists of nodes W2 (representing cave Gradišnica) and W3 (representing cave Gašpinova jama), connected by a conduit C1 and an overflow channel C2 at a position 20 m higher. The left node gets an inflow of simple trapezoidal shape, while the right node is connected to the first node and outfall. Detailed descriptions of all elements in the model are presented in Appendix C. Figure 5.32 shows the evolution of heads and flow rates in all conduits:

1. During the rise, the head drop between both nodes increases with an increase of flow (Figs. 5.31 and 5.32). The increase of the head is a result of increased friction in the connecting conduit.
2. When the overflow channel is reached, it takes the excess of water, keeping the head at node W2 (Gradišnica) almost constant (Figs. 5.28 and 5.30), but the inflow at node W3 (Gašpinova jama) continues to raise the head there (Figs. 5.28,

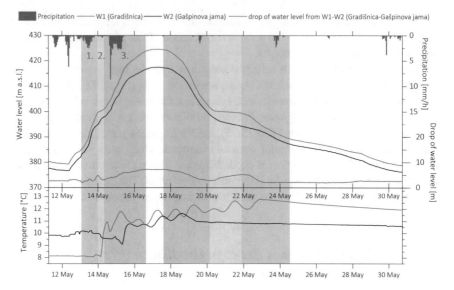

Fig. 5.28 Uniform fluctuation of water level in W2 (Gradišnica) and W3 (Gašpinova jama) during a high water event in May 2016. Purple line indicates slightly increased drop of the water level when water level is higher. The green and dark blue belts denote a more rapid change in water level in W2 (Gradišnica), and the pale blue belt denotes a more rapid change in W3 (Gašpinova jama)

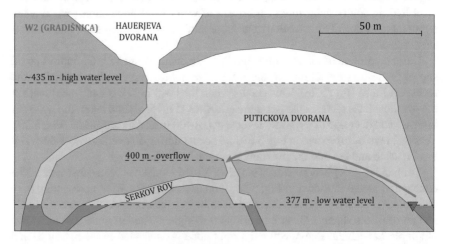

Fig. 5.29 The lowest part of W2 (Gradišnica) with two main chambers: Putickova dvorana and passage Šerkov rov into which the water from Putickova dvorana overflows (Cave data from Cave Register [5])

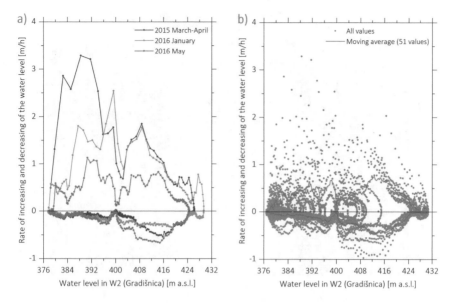

Fig. 5.30 Rate of water level increase and decrease in W2 (Gradišnica) as a function of the absolute water level: **a** three selected flood events; **b** all flood events combined

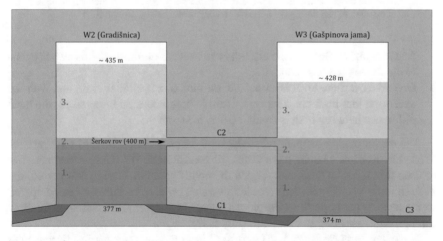

Fig. 5.31 Conceptual model of the assumed distribution of passages that are connecting W2 (Gradišnica) and W3 (Gašpinova jama). Both caves are presumably connected with overflow passage Šerkov rov resulting in inflections on the water level hydrographs. The presented conditions were used in the construction of the SWMM model. Numbers represent phases visible on water level hydrographs on Figs. 5.28 and 5.32

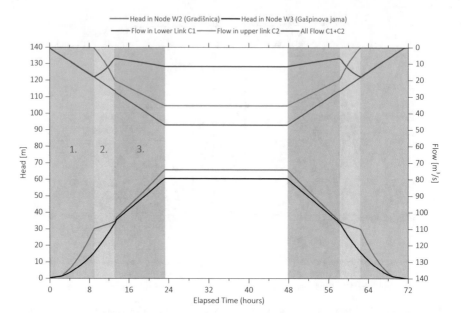

Fig. 5.32 Simulation of a high water event constructed in SWMM. The model conditions, presented in Fig. 5.31, were tested with a simple trapezoidal recharge. The test confirmed the changes to the water level dynamics seen during the actual high water event shown on Fig. 5.28

5.31, and 5.32). Because the difference in heads is diminished, this also happens to the flow in the base conduit.

3. Once the overflow level is surpassed, the situation is similar to before, with the exception that both conduits now connect both nodes and, therefore, the head difference between both is smaller (Figs. 5.31 and 5.32).

The mirror picture happens when the level drops to the position of the overflow. When the overflow channel loses water, the head at node W3 (Gašpinova jama) has to drop first to make the flow between the nodes possible. During that period, the head in node W2 (Gradišnica) recedes slowly. This drops relatively fast compared to the head at the node W3 (Gašpinova jama) (Fig. 5.32). This increases the flow through the connecting base conduit (Fig. 5.32).

Water temperature (Fig. 5.28) and SEC in both caves are similar during high water periods, when the water within the system is well mixed. During low water periods, the characteristics of both parameters evidently differ, which is the result of differences in the source of the water and in the geometry of the caves. The known part of W2 (Gradišnica) is located in a forested region, whereas most of the known parts of W3 (Gašpinova jama) lie below the town of Logatec. During periods of low water, SEC in W3 (Gašpinova jama) is much higher than in W2 (Gradišnica). This could possibly be due to more polluted water coming from the populated area towards the observation point in the sump. During low water stage, the temperature of the water in W2 (Gradišnica) is much colder (up to 4 °C) than in cave W3 (Gašpinova jama). The main reason for this is that W2 (Gradišnica) is exposed to the surface and behaves like a climatic cold trap [6] with an active winter air convection cell bringing cold air deep into the cave. This makes such caves colder than on average. The active flow bypasses the sensor only during high waters, while most of the time it is just below a stagnant water layer in contact with cave air.

5.2.5   The Broader Picture: Hydraulic Relations Between the Northern Ponors of Planinsko Polje (P2 and P3) and Caves Najdena Jama (W1), Gradišnica (W2), and Gašpinova Jama (W3)

In this section, the partial models described above are combined into a model of the entire system between the northern group of ponors on Planinsko Polje (P2 and P3) and W3 (Gašpinova jama). A conceptual model is presented in Fig. 5.33. As an inflow into the system, an event from May 2016 was taken.

The model consists of three nodes (W1, W2, and W3, representing caves Najdena jama, Gradišnica, and Gašpinova jama) and two inflows coming from the polje (P2 and P3, representing ponor zones Pod stenami and Škofov lom) (Fig. 5.33). P2, which is positioned lower, directly leads to the node W2 through conduit C1, while P3, which is positioned higher, leads through conduit C2 to the node W1. This node passes water further towards nodes W2 and W3. Nodes W1 and W2 are connected with conduit C3 and overflow passage C4 at 10 m, while nodes W2 and W3 are

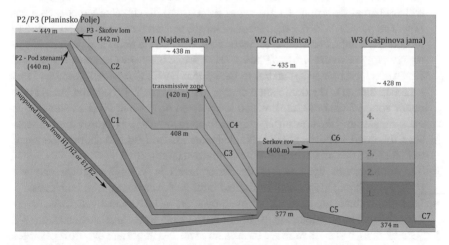

Fig. 5.33 Conceptual model of tested aquifer. Distribution of inflow, nodes, and conduits simulate possible characteristics of the system between the northern ponors of Planinsko Polje (P2 and P3) and W3 (Gašpinova jama). Numbers denote different phases of water level response also marked on Fig. 5.34

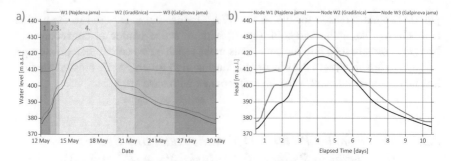

Fig. 5.34 a Water level hydrograph of actual event from May 2016, where different colours denote individual phases of water level response; **b** water level hydrograph with simulated high water event in the tested aquifer, shown also on Fig. 5.33

connected with conduit C5 and overflow passage C6 at 23 m. Downstream of W3 there is limited outflow through conduit C7 (Fig. 5.33).

Simulated hydrographs of water level dynamics in the model (Fig. 5.34b) matches well with the field measurements (Fig. 5.34a). A comparison of parameters, such as absolute water level, range of water level increase, position of inflection points, and the period of response at the specific location, are shown on both plots in Fig. 5.34. The modelling, therefore, supports the inferred mechanisms described in the previous sections.

New findings related to the studied system can be summarized with the following assertions:

- The connections between different locations, different input points, and different recharge areas determine the timing of response to floods recorded at each location. The area of W2-W3 (Gradišnica-Gašpinova jama) responds quickly due to its good hydraulic connection to ponor zone P2 (Pod stenami) and presumably to the recharge from the Hrušica Plateau (H1 [Veliko brezno v Grudnovi dolini] and H2 [Andrejevo brezno 1]) and/or to the system E1-E2 (Logarček-Vetrovna jama pri Laški kukavi). The dominant recharge to the area of W1 (Najdena jama) is the ponor zone P3 (Škofov lom), which is activated only when P2 (Pod stenami) is back-flooded and the flow on the polje is diverted towards P3 (Škofov lom). Therefore, the flood response at (W1) Najdena jama is delayed and shorter compared to that at W2 (Gradišnica) and W3 (Gašpinova jama).
- The simultaneous dynamics of the water levels in W2 (Gradišnica) and W3 (Gašpinova jama) are a result of limited outflow downstream from both caves.
- Inflection points are a result of overflows. In W1 (Najdena jama), an overflow channel is positioned at 10 m above the base water level, whereas, in W2 (Gradišnica), it is at 23 m above base water level. No important overflow passages have been identified beyond W3 (Gašpinova jama). The inflection point recorded in this cave is caused by an overflow between W2 (Gradišnica) and W3 (Gašpinova jama).

- Overflows cause a more rapid increase in the water levels at downstream locations. Overflow in W1 (Najdena jama) results in a more rapid increase of water level in W2 (Gradišnica), while overflow in W2 (Gradišnica) diminishes the difference in the water level towards W3 (Gašpinova jama).
- The volume of storage units and conduits play an important role on water level dynamics. A larger volume results in a slower increase and even in slower recession, resulting in an asymmetrical water level hydrograph. Storage is expected to be important in the region of caves W2 (Gradišnica) and W3 (Gašpinova jama).

5.3 System Related to the Hrušica Plateau

Two caves with permanent or ephemeral water were observed at the rim of the Hrušica Plateau—H1 (Veliko brezno v Grudnovi dolini) and H2 (Andrejevo brezno 1). They are located southwest from the Idrija Fault Zone (IFZ), which makes them relatively isolated compared to the other studied caves, located between Planinsko Polje and the springs of the Ljubljanica River (Fig. 5.35). The expected recharge area of these caves is the Hrušica Plateau and possibly the region of Hotenjski Ravnik.

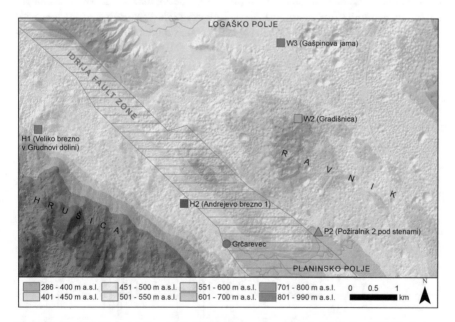

Fig. 5.35 Position of observation points H1 (Veliko brezno v Grudnovi dolini) and H2 (Andrejevo Brezno 1). Estavelles near Grčarevec at the northwestern rim of Planinsko Polje are marked with a red circle (DEM data from ARSO [2]; Cave data from Cave Register [5])

5.3.1   Groundwater Dynamics in the Cave Veliko Brezno v Grudnovi Dolini (H1)

Morphologically, the known part of H1 (Veliko Brezno v Grudnovi dolini) is a simple 90 m deep vertical shaft [5]. The cave has a stagnant low water level at 436 m a.s.l. Divers have explored the underwater part of the shaft to the depth of 30 m. During the highest water level, the cave acts as a spring.

The water level hydrograph shows a fast rise and recession for all events. Rise usually lasts between 6 and 20 h, while recession less than 3 days (Fig. 5.38). The highest recorded rise was 66 m. A typical event is shown Fig. 5.38a. The water level hydrograph shows a steep rise (on average 2–4 m/h, maximal >10 m/h) and recession (on average 0.6 m/h, maximal 1.2 m/h) from and back to the stagnant base level. The recession curve is concave, meaning that the rate of head drop increases towards the end of recession. The recession curve abruptly deflects to a straight line at 436 m a.s.l.

The water level hydrograph (Fig. 5.36) shows an extremely stable low water position at 436 m a.s.l., with a rapid response and sharp deflection to the stable state during recession. Such behaviour is typical for locations which are reached only by high waters, as the rest of the time, air pressure is being measured, which is characterised by a straight line on the water level hydrograph. Envisaging many scenarios and testing them with SWMM models, a conceptual model shown in Fig. 5.37, was completed.

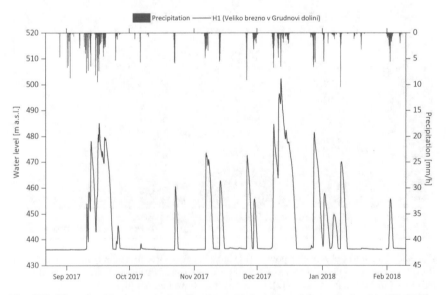

Fig. 5.36 Water level hydrograph showing the rapid dynamics of the water level in H1 (Veliko brezno v Grudnovi dolini) after each rain event

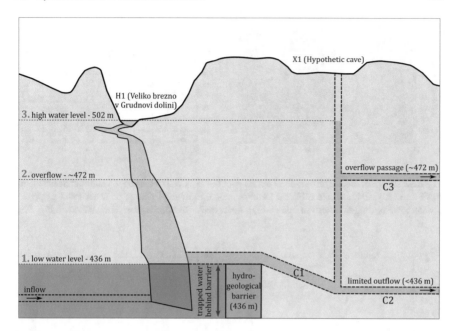

Fig. 5.37 Schematic view of possible cave (HX) that causes back-flooding in cave Veliko brezno v Grudnovi dolini (H1)

The part of the aquifer to which H1 (Veliko brezno v Grudnovi dolini) belongs is dammed with a low, permeable barrier on the northeast, which makes for a stable water level at 436 m a.s.l. The most plausible candidate is surely the Idrija Fault Zone. The base level on the opposite side of the barrier is unknown, but it is above the base level in W2 (Gradišnica) at 377 m a.s.l., possibly even above W1 (Najdena jama) at 410 m a.s.l., and below the base level in the area of H1 (Veliko brezno v Grudnovi dolini) (Fig. 5.37). During the event, the water from Hrušica Plateau flows along a highly transmissive zone above the barrier. The system on the northeastern side of the barrier responses quickly; the level rises above the barrier and back-floods the region of H1 (Veliko brezno v Grudnovi dolini). The recession is also mainly controlled by the level in the northeastern side of the barrier. Once the level there falls below the barrier, the stage in H1 (Veliko brezno v Grudnovi dolini) remains constant (Figs. 5.36 and 5.38).

During high floods, there is a clear deflection in the water level hydrograph at about 470 m a.s.l. (Fig. 5.38a). This is also visible on Fig. 5.39a, b, showing the rate of water level increase and decrease in H1 (Veliko brezno v Grudnovi dolini). A peak at 472 m a.s.l. indicates an overflow. There is no overflow channel known in H1 (Veliko brezno v Grudnovi dolini), however, it is also possible that this is an overflow within an unknown cave, X1 (marked with dashed lines on Fig. 5.37), which is hydraulically well connected with H1 (Veliko brezno v Grudnovi dolini).

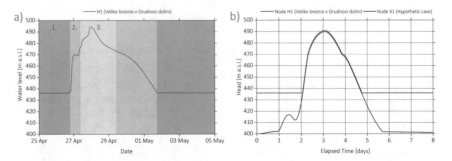

Fig. 5.38 a Water level hydrograph of actual event in H1 (Veliko brezno v Grudnovi dolini) from April and May 2017, where different colours denote individual phases of water level response; **b** water level hydrograph with simulated high water event in tested aquifer, shown also on Fig. 5.36

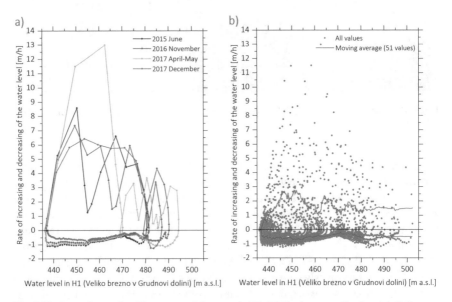

Fig. 5.39 Rate of water level increase and decrease in H1 (Veliko brezno v Grudnovi dolini); at an elevation around 472 m a.s.l., slower dynamics are visible: **a** four selected flood events; **b** all flood events combined

For demonstration, a simple SWMM model was built (Fig. 5.37), resulting in a similar response to the observed system (Fig. 5.38b). The model consists of two nodes (H1 and X1), representing caves Veliko brezno v Grudnovi dolini and the hypothetical cave located downstream of H1 (Fig. 5.37). Both caves are connected by a large channel C1, with an inlet position in H1 at 436 m a.s.l. (position of hydrogeological barrier) and an outlet position in X1 below this level (Fig. 5.37). X1 has a lower positioned outflow channel, C2, with limited outflow, and an upper positioned overflow channel, C3, at 472 m a.s.l. (Fig. 5.37). Detailed descriptions of all elements in the model are presented in Appendix E.

Different base levels at both caves, a good hydraulic connection between them, and limited outflow from the lower positioned cave result in back-flooding. Therefore, the water level hydrographs show simultaneous variation in the water level with different initial depths (Fig. 5.38b). At X1, full rising and recession limbs are observed, whereas recession at H1 stops abruptly with a straight line at 436 m a.s.l. (Fig. 5.38b), just as observed during actual events (Fig. 5.38a).

Water temperature variations in H1 (Veliko brezno v Grudnovi dolini) are small, between 8 and 9.25 °C (Fig. 5.40). During periods with low water levels, the temperature is stabilized at about 8.5 °C. This temperature reflects the equilibrium with the surrounding massif. This is also the stable cave-air temperature in nearby Andrejevo brezno 1 (H2) (Fig. 5.42). Although the temperatures do not change much, the dynamics are very peculiar, and show complex heat exchange between water, cave air, and massif.

During events, when the stage rises for more than 40 m, there is a clear drop of temperature to values below 8 °C (Fig. 5.40a). This probably indicates mixing (or through-flow) with colder water infiltrated at higher elevations in Hrušica Plateau.

Smaller events more or less show local mechanisms (Fig. 5.40b). Water at the position of the instrument is equilibrating with cave temperature and the equilibrium temperature of the massif. During small to medium events, the water rises from a few to 30 m. The instrument is exposed to the more equilibrated water, which was initially at the depth. The result is typically a slight decrease in temperature (Fig. 5.40b).

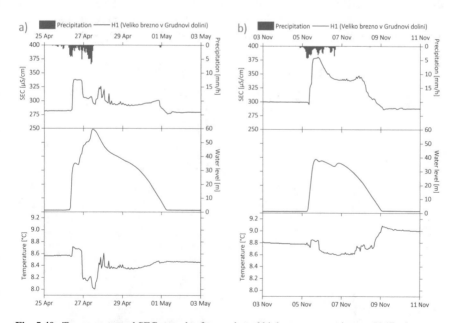

Fig. 5.40 Temperature and SEC records of two selected high water events in cave Veliko brezno v Grudnovi dolini (H1): **a** spring event from April 2017 with higher water level and lower temperature; **b** autumn event from November 2017 with lower water level and more intensive warming of water

Then, a peculiar temperature behaviour is recorded during most events. During recession, the temperature increases up to 9 °C (Fig. 5.40b). This happens in all seasons, even more characteristically in winter events. After recession, the water temperature at the position of the instrument equilibrates with the massif, unless disturbed by new event. The rise during recession is also recorded during small events. So far, there is no adequate model or mechanism which would explain this behaviour during water level recession (Fig. 5.40b).

Except during extreme hydrologic events, when the cave acts as a spring, the water level in the cave oscillates to accommodate increased pressure in the aquifer (the cave acts as a natural piezometer). The water at the position of the instrument is barely mixed with the event water (Fig. 5.40). The instrument is about 50 cm below the water surface. In a stable state, with no flow and mixing, the temperature must reflect that of the cave air temperature. During rise of the water level, therefore, the temperature does not change much and oscillates around 8.7 °C.

In H1 (Veliko brezno v Grudnovi dolini), the same holds for specific electrical conductivity (SEC). It has an annual cycle, with minimum SEC in winter and maximum in summer (Fig. 5.41). During flood events, the SEC increases; the highest recorded values of SEC are during high floods (Fig. 5.41). This indicates that the SEC increases with depth prior to the event and that during the rise of the water level, deeper and more mineralised water passes the instrument causing the increase in SEC. The opposite happens during recession. Several reasons can be foreseen for the stratification. The water at the surface may degas CO_2 into the cave atmosphere and precipitate calcite, leading to lower concentration and SEC close to the surface. The upper layer of water in the sump is also exposed to mixing with percolation

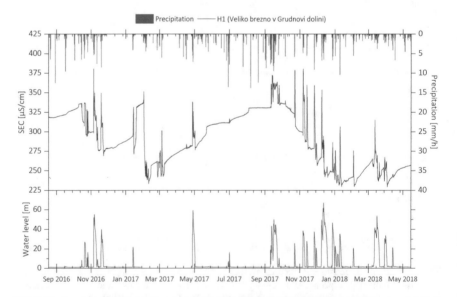

Fig. 5.41 Annual cycle of specific electrical conductivity in H1 (Veliko brezno v Grudnovi dolini)

water, which may have a lower concentration of dissolved calcium. The SEC also has an annual cycle with a similar phase as the typical annual cycles of soil CO_2 in the region [15]. This may support the idea that the inflow of percolating surface water with annual SEC cycles driven by soil CO_2 could play important role. However, it could also be that the CO_2 of the cave air has similar annual cycles. Additional measurements would be needed to give more precise answers.

5.3.2   *Groundwater Dynamics in the Cave Andrejevo Brezno 1 (H2)*

Like in H1 (Veliko brezno v Grudnovi dolini), H2 (Andrejevo brezno 1) also has a simple morphology with a 52 m deep vertical shaft and a lack of any known horizontal passages. During periods of low water, the cave has no access to the groundwater, while during high water events, more than half of the cave can be filled with water.

The water level hydrograph of H2 (Andrejevo brezno 1) (Fig. 5.42) shows relatively similar dynamics to H1 (Veliko brezno v Grudnovi dolini) (Fig. 5.36). Water level increases are rapid (from 12 to 30 h), whereas the decreases lasts from 2 to 10 days (Fig. 5.42). This is slightly slower than in H1 (Veliko brezno v Grudnovi dolini), but notably faster than in the caves located downstream from the northern and eastern ponors of Planinsko Polje. As the low water level is below the deepest known part of the cave, water level hydrographs do not show complete variation

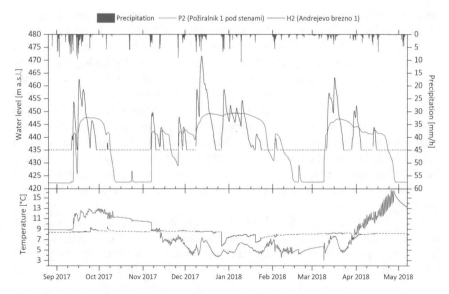

Fig. 5.42   Water level and temperature variation in cave Andrejevo brezno 1. Dashed line indicates position of the instrument, because the water level was below it

during high water events. A more detailed description of the water level dynamics in H2 (Andrejevo brezno) and its supposed relationships to Planinsko Polje, H1 (Veliko brezno v Grudnovi dolini), and W2 (Gradišnica) is in the following Sect. 5.3.3.

During most of the year, only the temperature of the cave air in H2 (Andrejevo brenzo 1) is registered (Fig. 5.42). This is stabilized at around 8.5 °C, which represents the temperature of the massif. Water temperature is, therefore, known only for high water periods, when the water level is high enough to flood the observation point. Because there is no active flow through the cave, the event water mainly bypasses it. The temperature of the water is, therefore, equilibrated to the temperature of the cave environment, so that it oscillates around 8.5 °C (Fig. 5.42). During particular situations, sudden changes of temperature are registered. They indicate a short period of water mixing, which is most likely the result of simultaneous discharging from both Planinsko Polje and Hrušica Plateau (Fig. 5.42).

The water level and temperature dynamics at H2 (Andrejevo brezno 1) can be discussed in relation to the broader picture, which includes flooding in Planinsko Polje, the dynamics in the region of H2 (Andrejevo brezno 1), and the role of the Idrija Fault Zone. Such discussion is given in the following Sect. 5.3.3.

5.3.3 Hydraulic Relations Between Veliko Brezno v Grudnovi Dolini (H1), Andrejevo Brezno 1 (H2), Planinsko Polje (P2) and Gradišnica (W2)

As mentioned previously, caves H1 (Veliko brezno v Grudnovi dolini) and H2 (Andrejevo brezno 1) are located southwest of the Idrija Fault Zone (IFZ), which presents a relatively strong hydrogeological barrier. The interaction of Planinsko Polje with an aquifer at its northwestern border (with H1—Veliko brezno v Grudnovi dolini and H2—Andrejevo brezno 1) was, until this study, poorly known. During flood events, the water level in the region of H1 (Veliko brezno v Grudnovi dolini) and H2 (Andrejevo brezno 1) responds quickly, much faster than the flooding in Planinsko Polje. This results in activation of numerous springs at the northwestern border of the polje, below the village of Grčarevec. On the other hand, the recession at H1 (Veliko brezno v Grudnovi dolini) and H2 (Andrejevo brezno 1) also precedes the flood recession in the polje. Therefore, the water from the flooded polje drains into the aquifer with H1 and H2 (Veliko brezno v Grudnovi dolini and Andrejevo brezno 1).

In the following paragraphs, different stages of flood events are presented (blue lines on Fig. 5.43). Records of the water level in W2 (cave Gradišnica), which is considered representative of the regional water level around the town of Logatec, is also shown. The area of W2 (Gradišnica) is a possible collector of all water from

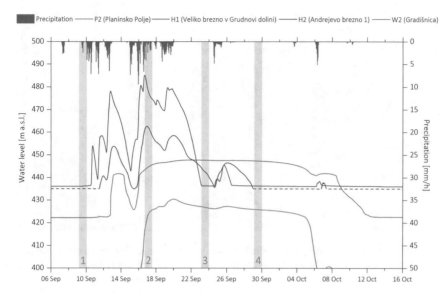

Fig. 5.43   Variations in the water level on both sides of IFZ (Idrija Fault Zone) during a high water event in autumn 2017. Note the periods when the water level is higher on the western side (H1 [Veliko brezno v Grudnovi dolini] and H2 [Andrejevo brezno 1]), and periods when the level is higher on Planinsko Polje. For a comparison of a downstream location, Gradišnica (W2) is added. Dashed line denotes periods, when water was below the position of instrument. Blue lines denote situations with a different distribution of the water level. These situations were tested with a SWMM model and are interpreted in following sections

Planinsko Polje, H1 (Veliko brezno v Grudnovi dolini), and H2 (cave Andrejevo brezno 1).

Situation 1: Low Water on the Southwest and Northeast

The groundwater dynamics during periods of low water is not well known (line number 1 on Fig. 5.43). The stage of the water level during this period is well known only in W2 (Gradišnica—377 m a.s.l.) and H1 (Veliko brezno v Grudnovi dolini—436.5 m a.s.l.). At these two caves, low water level is quite stable and most likely controlled by hydrogeological barriers (Fig. 5.44). In P2 (ponor Požiralnik 2 pod stenami), the water level is positioned at 422.5 m a.s.l., however, it is not known if this is the actual local water table or trapped water. In H2 (Andrejevo brezno 1), there is no access to the water during low water periods, so the water level is positioned below the lowest know point of the cave (435 m a.s.l.) (Fig. 5.44).

The water level in caves positioned southwest from the Idrija Fault Zone is higher than the water level on Planinsko Polje, but the most probable direction of the groundwater flow is towards the region of W2 (Gradišnica) (Fig. 5.44) and W3 (Gašpinova jama). In this case, the Idrija Fault Zone represents a hydrogeological barrier, which keeps the water level on the southwest at 436 m a.s.l., but has a highly transmissive zone above this level, which allows water from the Hrušica Plateau and Hotenjski

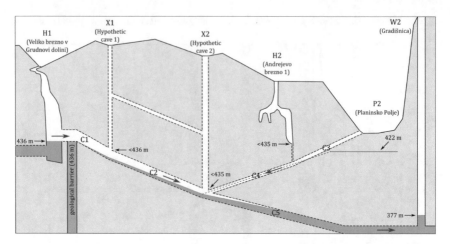

Fig. 5.44 Hydrological situation in selected caves at low water level. General direction of flow is towards W3 (Gradišnica). The size of the arrows indicates a small amount of flow

Ravnik to flow freely towards the region of W2 (Gradišnica) and W3 (Gašpinova jama) (Fig. 5.44).

Situation 2: High Water in the Southwest, Increase of Water in the Northeast
During flood events (line number 2 on Fig. 5.43), the recharge from the Hrušica Plateau is rapid and conveyed through a high transmissivity zone above 436 m a.s.l. The general direction of the flow is towards the northeast (W2—Gradišnica) (Fig. 5.45), but because of limited outflow through C2 and C5 (Fig. 5.45), it back-floods above the barrier, which is recorded in the response in H1 (Veliko brezno v Grudnovi dolini). The flooding in the southwestern region is also recorded in H2 (Andrejevo brezno 1) (Fig. 5.45). During extreme events, the level in the region of H2 (Andrejevo brezno 1) is above Planinsko Polje and the water discharges to the polje through springs near Grčarevec at its northwestern border (Figs. 5.45 and 5.46). They contribute a relatively important amount of water (several m^3/s).

During the observation period, the increase of the water level in H1 (Veliko brezno v Grudnovi dolini) was up to 66 m and at the highest stage, a doline with an entrance was also flooded (Fig. 5.47). Other dolines in the vicinity were partially flooded as well. The increase of the water level in H2 (Andrejevo brezno 1) was at least 31 m high.

Situations 3 and 4: Decrease in the Southwest, High Water in the Northeast
The recession starts rapidly in the region of H1 (Veliko brezno v Grudnovi dolini) and H2 (Andrejevo brezno 1). Once the water level behind the barrier falls below the level of the barrier at 436 m a.s.l., the level in H1 (Veliko brezno v Grudnovi dolini) stays at this level (Fig. 5.48).

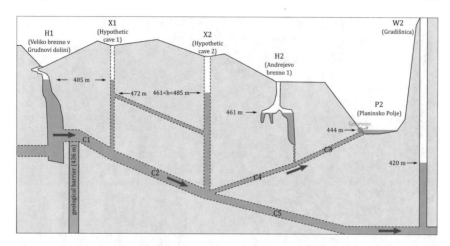

Fig. 5.45 Hydrological situation in selected caves at the beginning of the high water event. High water level in the Hrušica Plateau activates the springs near Grčarevec. The size of the arrows indicates the amount of flow

Fig. 5.46 Active estavelles near Grčarevec acting as springs (*Photos* M. Blatnik)

Due to its large recharge area and storage, the floods on Planinsko Polje last longer than the flood events caused by the recharge from the Hrušica Plateau. When the level in the region of H2 (Andrejevo brezno 1) drops below the polje, the springs near Grčarevec turn into ponors (Fig. 5.48). The recharge from Planinsko Polje causes slower recession in H2 (Andrejevo brezno 1) than it occurs in H1 (Veliko brezno v

Fig. 5.47 H1 (Veliko brezno v Grudnovi dolini) has a quick recession of the water level: **a** melted snow denotes the highest water level during the observation period (12. 12. 2017 at 10:00); 6 h later, the water level was 4.5 m lower and covered the lowest part of doline with the cave entrance (*Photo* A. Mihevc); **b** 13. 12. 2017 at 14:30 the water level was 16.5 m below the highest level and was positioned on the top of the main shaft in the cave, shown on the right picture (*Photo* M. Blatnik)

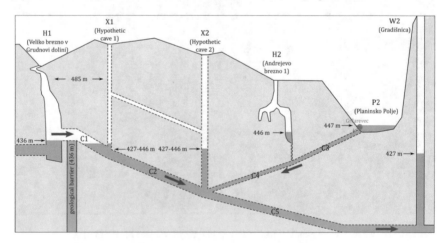

Fig. 5.48 Hydrological situation in selected caves during recession, when H1 (Veliko brezno v Grudnovi dolini) has the lowest water level. The size of the arrows indicates the amount of flow

Grudnovi dolini) (Figs. 5.43 and 5.48). For some time, the water level in H2 can be higher than in H1 (line number 3 on Figs. 5.43 and 5.48).

When the water level drops below the lowest known part of the cave (435 m a.s.l.), only atmospheric pressure is measured, but inflow from the polje keeps the level close to the position of the instrument (Fig. 5.50). This can be seen on Fig. 5.49, where a small additional rain event during the recession stage causes a minimal rise of the water level in H1 (Veliko brezno v Grudnovi dolini), while in H2 (Andrejevo brezno 1), the rise is distinct (about 7 m). A small change in H1 (Veliko brezno v Grudnovi dolini) indicates an almost free flow of water over the barrier, while additional flow from Planinsko Polje results in back-flooding in H2 (Andrejevo brezno 1). Even more interesting is the temperature at H2 (Andrejevo brezno 1), which abruptly drops below normal air or water temperature prior to the event (Fig. 5.49). It clearly indicates that the water from Planinsko Polje flows below H2 (Andrejevo brezno 1) and is forced to fill the lower part of the cave during the event.

During recession, all water flows towards the region of W2 (Gradišnica) and W3 (Gašpinova jama). At some point (line number 4 on Fig. 5.43), when H2 (Andrejevo brezno 1) is dry (water level below 435 m a.s.l.), W2 (Gradišnica), located downstream, has the highest water level (about 425 m a.s.l.). During this period, the drop of the water level is small (less than 10 m) and the gradient is less than 0.003 (Fig. 5.48). Towards W3 (Gašpinova jama) the drop of the water level and gradient are slightly larger (15 m; 0.005).

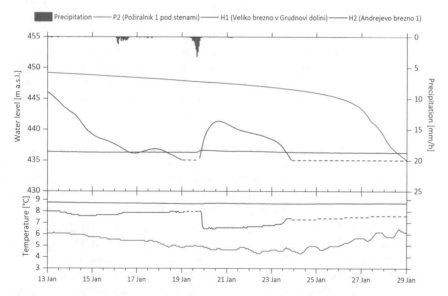

Fig. 5.49 Additional rain event on 20th January 2018 that caused almost no change of the water level in H1 (Veliko brezno v Grudnovi dolini), but a distinct rise in H2 (Andrejevo brezno 1), because of additional inflow from Planinsko Polje (P2)

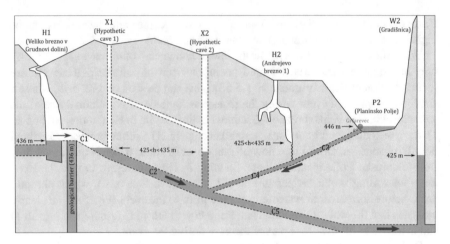

Fig. 5.50 Hydrological situation in selected caves during recession, when cave Andrejevo brezno 1 has the lowest water level. The size of the arrows indicates the amount of flow

To simulate the behaviour, a SWMM model has been constructed. The modelling domain is shown on Figs. 5.44, 5.45, 5.48, 5.50. It consists of three known caves, H1 (Veliko Brezno v Grudnovi dolini), H2 (Andrejevo brezno 1), and W2 (Gradišnica), two hypothetical caves (X1 and X2), and inflow P2 (Planinsko Polje). H1 directly receives some amount of water (inflow from the Hrušica Plateau) and is further connected to the hypothetical cave X1 with a large conduit C1. The inlet elevation of C1 is at 436 m a.s.l. (representing overflow over the geological barrier), while the outlet is below this elevation. After X1, the outflow is limited by channel C2, which causes back-flooding in both caves H1 and X1, as described in Sect. 5.3.1. Located downstream from caves H1 and X1 is another cave, X2. H2 receives water from P2 and also from the Hrušica Plateau. Downstream from all described nodes is the hypothetical cave X2. That cave is also discharged from both sources and because of limited outflow through channel C5, causes back-flooding in all upstream caves. All water continues to flow towards W2. Detailed descriptions of all elements in the model are presented in Appendix F.

The simulation model confirms that the contribution of water from Hrušica Plateau and Planinsko Polje changes over time (Fig. 5.51b). Rapid inflow from the Hrušica Plateau results in a rapid increase of the water level in H1 (Veliko brezno v Grudnovi dolini) that is caused by back-flooding. Back-flooding also causes a rapid increase of the water level in H2 (Andrejevo brezno 1) (Fig. 5.51b). The water level increases to a higher position than in P2 (Planinsko Polje), but because of the smaller diameter of the channel and some hydraulic gradient, it is smaller than in H1. At the same time, Planinsko Polje is flooding. When the inflow on Hrušica Plateau ceases, the water level in H1 rapidly decreases to its initial level (Fig. 5.51b). Decrease in H2 is slower because of additional inflow from P2 (flooded Planinsko Polje) and persists until the outflow becomes larger than the inflow (Fig. 5.51b). Figure 5.51a presents the series of the actual high water event, which agrees with the results of the simulated model.

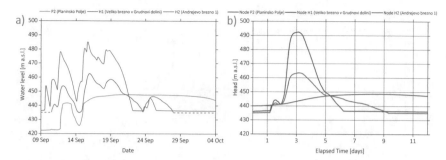

Fig. 5.51 Relation between water level in P2 (Planinsko Polje) and caves H1 (Veliko brezno v Grudnovi dolini) and H2 (Andrejevo brezno 1) during high water: **a** the actual measurements from September 2017; **b** modelled dynamics, according to the settings shown on Figs. 5.44, 5.45, 5.48 and 5.50, and the actual measurements during the rain event from May 2016

5.4 Temperature and Specific Electrical Conductivity (SEC) Records at the Springs of the Ljubljanica River

Temperature and SEC at three major springs of the Ljubljanica River were recorded from the 16th of September 2016 to the 9th of May 2018. They are the springs Močilnik (belonging to the Mala Ljubljanica River), springs Retovje (a tributary to the Velika Ljubljanica River), and one of the springs of the Bistra River. All studied springs are perennial, with each representing a different group of springs feeding the main tributaries of the Ljubljanica River.

These records were compared to the records from caves located upstream and, although there were some limitations, transit times of the groundwater flow were calculated and the possible extent of the recharge area was identified.

5.4.1 Temperature Characteristics

All studied springs show similar temperature dynamics. The temperatures show annual cycles with distinct disturbances during high water events. The rest of the time, slow changes are present and they are controlled by the seasons of the year.

From September 2016 to May 2018, the annual temperature varied from 5.7 to 13.8 °C (Fig. 5.52), where the general temperature increased from the western positioned springs toward the eastern springs. During high water events, temperature differences were small, whereas during low water conditions, differences were up to 2 °C. Because of the 11 km long groundwater flow and efficient heat exchange between the water and cave walls, the diurnal temperature oscillations from the surface stream of the Unica River to the springs of the Ljubljanica River are fully damped (Fig. 5.52).

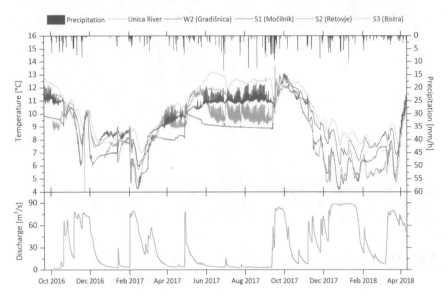

Fig. 5.52 Seasonal temperature dynamics of the Ljubljanica River springs, the cave Gradišnica located upstream, and the Unica River

Besides the generally similar dynamics of temperature variation at all springs, some differences are present. At springs Močilnik, diurnal temperature oscillations can occur (Fig. 5.52). There, the location of the instrument is about 100 m from the spring, which is far enough for heat exchange to occur, causing distinct diurnal temperature dynamics during low-flow conditions.

Of all of the studied springs, Bistra has a distinctly different temperature profile (Fig. 5.52). There could be multiple reasons for this, but the most plausible is due to a higher contribution of water coming from karst poljes, particularly from Cerkniško Polje. Močilnik and Retovje are dominantly fed by waters from Planinsko Polje (the Unica River), whereas Bistra is fed not only from Planinsko Polje, but from Cerkniško Polje (the Stržen River) as well. The biggest differences occur during low water conditions, when water flowing over the poljes is most susceptible to heat exchange with the air. During the summer, Bistra is the warmest (Fig. 5.53b), while in winter, it becomes the coldest (Fig. 5.53a). Additionally, during periods of low water, Bistra has the highest discharge among all three studied springs (Fig. 5.53). According to dye tracing in the 1970s (Gospodarič and Habič 1976), Bistra receives most of its water from Cerkniško Polje, which has a much higher temperature variability than the Unica River. This is also reflected in the observations of Bistra given above.

The annual temperature variations at the springs correlate nicely to those of the Unica River throughout the entire year. The temperature signal of the Unica River is damped and delayed along the flow. Diurnal temperature signals are, as expected, not registered at the springs (Fig. 5.53). Spring temperatures show a close relation to cave temperature only during high water conditions, when active flow is present

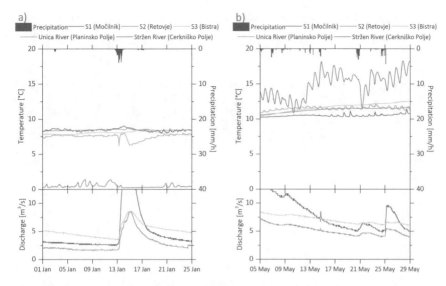

Fig. 5.53 Temperature dynamics of the springs of the Ljubljanica River during low water conditions, when the greatest amount of water is drained from the poljes: **a** winter situation from January 2017, when the springs of Bistra are the coldest; **b** spring and summer situation from May 2017, when the springs of Bistra are the warmest

at the instrument locations within the caves. In comparison to the springs, diurnal temperature oscillations are preserved (Fig. 5.53) in caves due to their proximity to the points of recharge.

5.4.2 Characteristics of Specific Electrical Conductivity (SEC)

Specific electrical conductivity (SEC) measurements also show a similar pattern at all studied springs. It is strongly controlled by the hydrological situation (Fig. 5.54). The annual amplitude of SEC at the springs varies from 300 to 475 μS/cm, where both extremes are related to high water events (Fig. 5.54).

During high water events, the SEC dynamics rapidly fluctuate, indicating the mixing of different waters. The large amount of event water diminishes the SEC at all springs (Fig. 5.54). During the recession, the SEC increases exponentially (Fig. 5.54). The fast flow component diminishes, while the contribution of chemically well equilibrated slow flow, autogenic recharge, and stored water increases. This is characteristic for all observed springs, but not in most caves located upstream (for example Gradišnica) with stagnant water during low flow (Fig. 5.54).

The drop in SEC during flood events is much larger at springs Močilnik and Retovje, where the dynamics of SEC closely resemble that of the Unica River

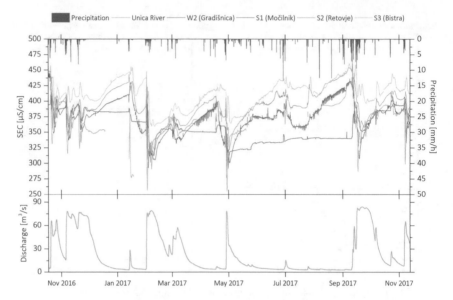

Fig. 5.54 Specific electrical conductivity dynamics of the Ljubljanica River springs, the cave Gradišnica located upstream, and the Unica River during selected high water events

(Fig. 5.54). In Bistra, however, its contribution is lower and, therefore, the drop of SEC is lower. The SEC in Bistra is generally from 10 to 40 μS/cm higher than in Močilnik and Retovje, which may indicate a higher contribution of slowly draining water coming from the karst poljes, the same reason that these springs are a bit warmer (Figs. 5.53 and 5.54).

5.5 Regional Position and Dynamics of the Water Level

5.5.1 General Characteristics of the Regional Water Level

The karst aquifer in this study extends between the Ljubljana Basin at 290 m a.s.l. and Planinsko Polje with a floor at 440 m a.s.l. A 150 m drop over an approximately 10 km straight-line distance gives an average gradient around 0.015. These two altitudes also represent the approximate position of the highest and lowest water table in the area.

Over 3.5 years of observation, there were many events where an increase in the water level occurred (Fig. 5.55). Most of them were relatively short (several days long) with a slightly increased water level. There were 15 events when Planinsko Polje was flooded. Such periods lasted from about one week up to three months (Fig. 5.55).

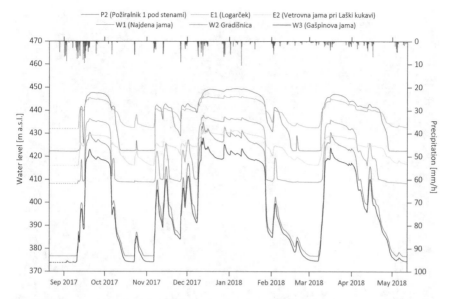

Fig. 5.55 Plot showing the high frequency of high water events from autumn 2017 to spring 2018. Plot presents absolute water level at selected observation points. Dashed lines indicate periods when water level was below the instrument

During floods, the water level in the aquifer reaches its upper limit. In the studied karst massif, the rise of the water level was up to 58 m in W2 (Gradišnica) (Figs. 5.55, 5.56, 5.57, and Table 5.2). An even higher response (66 m) was recorded in the cave H1 (Veliko brezno v Grudnovi dolini) (Table 5.2) on the southern side of the Idrija Fault Zone. For some of the studied caves, the water level amplitude is not known because the water level is not accessible during low water (Table 5.2).

Simple interpolations of the water table during the lowest and the highest water level indicate different hydraulic gradients (Fig. 5.57). The surface of the low water table has a concave shape with a high gradient (high density of isolines) in part of the aquifer close to Planinsko Polje and a smaller gradient downstream from the cave W3 (Gašpinova jama) (Fig. 5.57a). During high water level conditions, the surface of the water table is convex (Fig. 5.57b). The lowest gradient (low density of isolines) is in the area close to Planinsko Polje, whereas downstream the gradient is larger.

The interpolations are very general and do not include the influence of hydro-geological barriers and flow in different porous media. Additionally, the density of observation points is scarce, particularly in the northern part of the study area.

Fig. 5.56 Cave Gradišnica at the highest water level (*Photo* M. Blatnik)

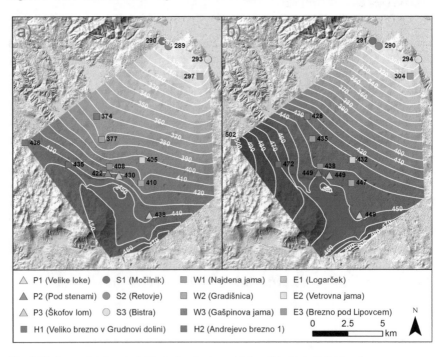

Fig. 5.57 Interpolated water table during the lowest (**a**) and the highest (**b**) level

Table 5.2 List of observed caves with the highest and lowest recorded water levels and their total amplitude

Name of observation point	Low water level	High water level	Amplitude
P1 (ponor Velike loke)	Below 440.7	448.4	>7.7
P2 (ponor Požiralnik 2 pod stenami)	Below 430.0	448.4	>18.4
P2 (ponor Požiralnik 1 pod stenami)	422.2	449.5	27.3
P3 (ponor Požiralnik 1 v Škofovem lomu)	Below 430.4	447.1	>16.7
W1 (Najdena jama)	408.0	438.0	30.0
W2 (Gradišnica)	376.8	435.0	58.2
W3 (Gašpinova jama)	373.9	428.0	54.1
E1 (Logarček)	410.0	446.6	36.5
E2 (Vetrovna jama pri Laški kukavi)	416.0	432.9	17.0
E3 (Brezno pod Lipovcem)	Below 297.1	304.1	>7.0
H1 (Veliko brezno v Grudnovi dolini)	436.3	502.3	66.0
H2 (Andrejevo brezno 1)	Below 435.0	471.6	>36.6

Note that in some caves there is no water during periods of low water, so low water level and amplitude are unknown

5.5.2 Typical Evolution of Flood Event

High water events have various durations. These depend on precipitation events (rain intensity, distribution, and duration) and the initial hydrological conditions (dry or wet period, season with high or low evapotranspiration). There is also a difference in the response and duration of different phases at selected observation points, which is controlled by their position within the aquifer, local geometry, and recharge. However, most major flood events go through the following stages of rising and recession (Figs. 5.56 and 5.58).

Rising limbs:

1. **Pre-event low water position**. The Unica River discharge is low, ponor zones inactive, and groundwater level at its lowest stage (Figs. 5.58 and 5.60a).
2. **Quick response in the Hrušica Plateau (H1-H2 block)**. The first evident response to the rain event occurs in the caves associated with the recharge from the Hrušica Plateau (H1 and H2) (Figs. 5.58 and 5.60b). Because of outflow is limited, back-flooding of H1 (Veliko brezno v Grudnovi dolini) and H2 (Andrejevo brezno 1) occurs. This surely triggers the filling of block W2-W3 (Gradišnica-Gašpinova jama).
3. **Outflow through the eastern ponors of Planinsko Polje and the response in Logarček-Vetrovna jama pri Laški kukavi system (E1-E2 block)**. Increased flow of the Unica River first reaches the eastern ponor zone of Planinsko Polje, which results in the filling of block E1-E2 (Logarček-Vetrovna jama pri Laški kukavi) (Figs. 5.58 and 5.60b).

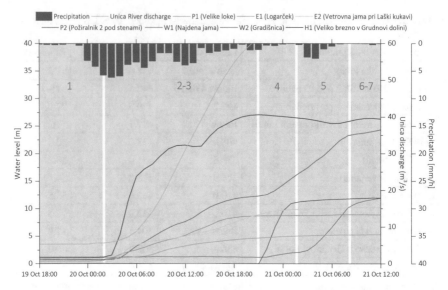

Fig. 5.58 Water level and discharge hydrograph showing the beginning of the high water event from October 2016

4. **Outflow through the ponor zone Pod stenami and additional rise in Gradišnica-Gašpinova jama system (W2-W3 block)**. With additional rising of the Unica River discharge, the group of ponors of Pod stenami on the northern border of Planinsko Polje activates and causes an even more rapid increase of the water level in block W2-W3 (Gradišnica-Gašpinova jama) (Fig. 5.58).
5. **Outflow through the ponor zone Škofov lom and response in the Najdena jama system (W1 block). Additional rise in W2-W3 block**. When outflow at ponor zone Pod stenami is surpassed, overflow towards ponor zone Škofov lom occurs (Fig. 5.58). Activation of these ponors results in the filling of block W1 (Najdena jama). This results in an additional increase of the water level in block W2-W3 (Gradišnica-Gašpinova jama) (Fig. 5.58).
6. **Flooding of the polje with recharge from springs on the southern side and springs below Grčarevec**. When the aquifer is partially filled and the outflow capacity of the ponors is surpassed by the discharge of the Unica River, Planinsko Polje starts to flood (Figs. 5.58 and 5.60c). Because of the higher groundwater level in the Hrušica Plateau, additional water comes to the polje through springs near the village of Grčarevec (Fig. 5.60b, c).
7. **Flooded polje**. Depending on rain frequency and intensity, Planinsko Polje can be flooded from several days to several months. During the 3.5 year long observation period, the longest period of flooding lasted more than three months (Fig. 5.55). Figure 5.60c shows a schematic view of the water level during the highest measured extent.

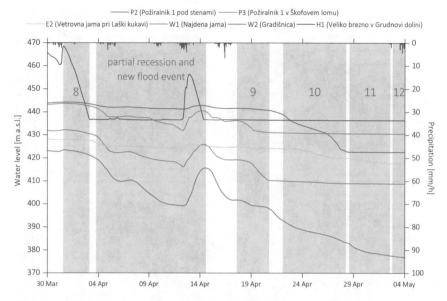

Fig. 5.59   Water level hydrograph showing recession of the high water event in April 2018

Recession:

8. **Fast recession in the Hrušica Plateau**. The recession is the fastest in block
   H1-H2 (Veliko brezno v Grudnovi dolini-Andrejevo brezno 1) (Figs. 5.59 and
   5.60d), where water level drops below the position of Planinsko Polje, and the
   springs near the village of Grčarevec turn into ponors (Fig. 5.60d) if the polje
   is flooded.

9. **Deactivation of the ponor zone Škofov lom and initial drop of water level in
   systems Najdena jama (W1 Block) and Gradišnica-Gašpinova jama (W2-
   W3 block)**. Recession in Planinsko Polje begins with the switching-off of
   Škofov lom, the highest positioned ponors (Fig. 5.59). This causes reduced
   inflow in the systems of blocks W1 (Najdena jama) and W2-W3 (Gradišnica-
   Gašpinova jama) and an initial drop in the water level hydrograph. When ponor
   zone P3 (Škofov lom) is completely deactivated, the water in block W1
   (Najdena jama) drops to the base level (Fig. 5.59).

10. **Deactivation of the ponor zone Pod stenami and the main drop in
    Gradišnica-Gašpinova jama system (W2-W3 block)**. With further recession
    on Planinsko Polje, ponor zone Pod stenami is deactivated (Fig. 5.59). Because
    the inflow fades out, the water level in the block W2-W3 (Gradišnica-Gašpinova
    jama) system also slowly decreases towards the base level (Fig. 5.59).

11. **Attenuation of the eastern ponors and the main drop in Logarček-Vetrovna
    jama pri Laški kukavi system (E1-E2 block)**. When the polje is no longer
    flooded and the flow of the Unica River is further decreased, the set of eastern
    ponors is deactivated. This results in a gradual decrease of the water level in the

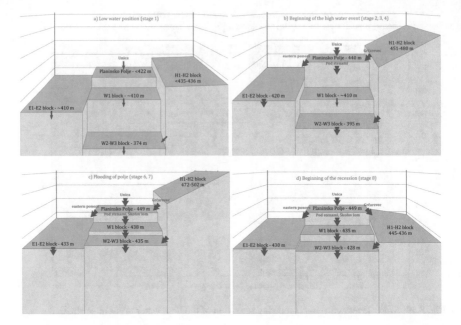

Fig. 5.60 Schematic view of the water level in determined blocks at various stages of the high water event. The size of the arrows indicates the estimated amount of flow. The view is from the north (front) towards the south (back)

    block E1-E2 (Logarček—Vetrovna jama pri Laški kukavi) system to the base level (Fig. 5.59).

12. **Recession to base level**. The Unica River discharge decreases to the lowest extent, the ponor zones are inactive, and the groundwater level is at its lowest stage (Figs. 5.59 and 5.60a).

Figure 5.60 schematically shows the distribution of groundwater level at different phases of high water events. Note that the step-drops between the blocks do not represent the physical realm. There are indications of low transmissivity zones between the blocks, however such discontinuities are not expected or even possible. The steps in the figure are kept for the clarity of presentation.

5.6   Regional Characteristics of Water Temperature and Specific Electrical Conductivity

5.6.1   General Characteristics of Water Temperature

The temperature of the water strongly depends on the local climate. The climate controls the temperature of surface streams through the air temperature, while groundwater temperature is controlled through the temperature of the rock [9]. The average air temperature of the Ljubljanica River recharge area varies from about 5 °C on the High Dinaric plateaus to about 10 °C in the town of Vrhnika on the border of the Ljubljana Basin [1].

Surface streams are exposed to seasonal and diurnal temperature forcing. Solar radiation and heat exchange with the air causes warming of the water during the day and cooling during the night [7]. Therefore, surface streams carry a distinct diurnal temperature signal, the amplitude of which depends on the weather (temperature, sky clearness, quantity, and duration of flow). When entering underground, the heat exchange between water and rock diminishes the amplitude of the temperature [8, 11]. As no diurnal forcing is present in the underground, the phase signal is almost "locked" along the flow and can thus be used for the estimation of transit times and apparent velocities between consecutive points of the flow system (see Sect. 5.6.5).

There are four observation points in the surface stream of the Unica River. Ponor P2 (Velike loke) is positioned in the ponor zone after 7 km of the surface flow. It shows clear diurnal temperature oscillations most of the time except for during very low flow conditions when the ponor is dry (Fig. 5.61). Ponor zones P2 (Pod stenami) and P3 (Škofov lom) are located on the northern side of the polje, after 17 km of the surface flow. These ponors are flooded only during medium to high water, when diurnal temperature oscillations are evident and larger than in P1 (Velike loke). The amplitude of the diurnal oscillations can reach up to 3.6 °C (Fig. 5.61 and Table 5.5).

Most observation points in the caves are in stagnant water during low flow and medium flow conditions. An exception is E2 (Vetrovna jama pri Laški kukavi), where the instrument is also in moving water during low flow conditions. In H2 (Andrejevo brezno 1) and E3 (Brezno pod Lipovcem), the water reaches the instruments only during floods (Fig. 5.61).

During low flow, the subsurface temperature of the stagnant water is equilibrating to the ambient temperature of the cave. Most observation points are relatively isolated in respect to the entrances, therefore the cave temperatures are considered a good indicator of the massif temperature (Fig. 5.61). An exception is W2 (Gradišnica), which is widely exposed to the surface through a shaft and inclined passage. There, a convection cell builds up in the winter and brings cold air deep into the cave. The cell is switched off during the summer, so the cave acts as a cold trap [6]. For this reason, the water temperature in W2 (Gradišnica) is generally colder (Fig. 5.61 and Table 5.3).

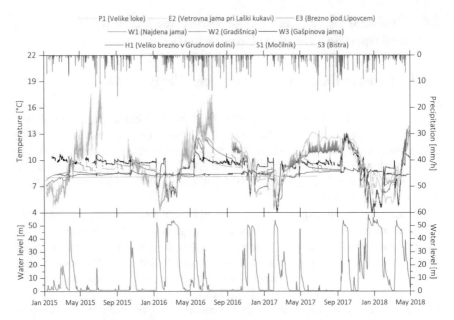

Fig. 5.61 Water temperatures at selected observation points from January 2015 to May 2018. Dashed lines and missing values indicate periods when the water level was below the instrument. Lower graph shows the water level

During flood events, several things may happen:

- In some caves (W1—Najdena jama, W2—Gradišnica, W3—Gašpinova jama, E1—Logarček), the flow is partially (or largely) diverted into the higher bypasses, so that the flow passes the observation points. In this situation, the temperature of the event flow with a possible diurnal character is observed (Fig. 5.61 and Table 5.5).
- During small events, and in some caves (H1—Veliko brezno v Grudnovi dolini, H2—Andrejevo brezno 1, E3—Brezno pod Lipovcem) during most of the events, the stagnant water more or less rises and drops with no presence of the active flow at the observation points. In this case, the temperature of the stagnant water is sensed (Fig. 5.61). This water may be partially mixed with event water if the active channel is close enough. In such situations, even if the diurnal signal is present in the active channels, it is not recorded by the instruments (Table 5.5).
- As stated above, only observation point E2 (Vetrovna jama pri Laški kukavi) is in the active stream. There, diurnal oscillations are observed during all medium and large events (Fig. 5.61 and Table 5.5).

At all three observed springs of the Ljubljanica River, relatively stable temperatures with a clear annual cycle were observed (Fig. 5.61). Sudden temperature changes were observed during flood events. Because of the long underground flow,

Table 5.3 Measured minimal, maximal, and mean temperature of water during selected periods

Name of observation point	Minimal temperature	Maximal temperature	Mean temperature	Mean temperature calculation period
Unica River (Hasberg)	4.20	13.70	8.92	Mar 2017–Mar 2018
P1 (ponor Velike loke)	4.15	19.05	9.25	Mar 2015–Mar 2016
P2 (ponor zone Pod stenami)	3.05	16.50	7.90	Mar 2015–Mar 2018
W1 (Najdena jama)	4.00	13.45	8.95	Mar 2015–Mar 2018
W2 (Gradišnica)	4.20	13.55	8.65	Mar 2015–Mar 2018
W3 (Gašpinova jama)	4.10	12.80	9.25	Mar 2015–Mar 2018
E1 (Logarček)	4.40	16.90	9.75	Mar 2015–Mar 2018
E2 (Vetrovna jama pri Laški kukavi)	4.50	16.25	9.12	Mar 2015–Mar 2018
E3 (Brezno pod Lipovcem)	8.20	8.50	8.30	Mar 2015–Mar 2017
H1 (Veliko brezno v Grudnovi dolini)	7.70	9.25	8.50	Mar 2015–Mar 2018
H2 (Andrejevo brezno 1)	5.75	9.45	8.15	Mar 2017–Mar 2018
S1 (springs Močilnik)	6.10	13.85	9.85	Mar 2017–Mar 2018
S2 (springs Retovje)	5.90	13.28	10.10	Mar 2017–Mar 2018
S3 (springs Bistra)	7.35	13.25	10.75	Mar 2017–Mar 2018

diurnal temperature oscillations are completely diminished (Table 5.5). The exception is the spring S1 (Močilnik), where the observation point is about 100 m downstream from the spring, so the local diurnal forcing at the spring is also recorded (Fig. 5.61). From west to east, temperatures generally increase due to a larger contribution of water from Cerkniško Polje (Table 5.3).

5.6.2 General Characteristics of Specific Electrical Conductivity

Specific electrical conductivity (SEC) of karst waters mainly depends on the concentration of dissolved Ca^{++} ions [14]. This is related to the hydrological conditions and lithology (carbonate vs. non-carbonate rocks) of the recharge area. Water in the aquifer is driven to a chemical equilibrium with the surrounding rock [14]. During low flow, the water in the aquifer is closer to the equilibrium, which results in a higher SEC. During flood events, the stored water mixes with a large amount of fresh non-equilibrated water, resulting in a distinct drop in SEC (Fig. 5.62). Pollution brings in foreign ions and increases conductivity. In the studied system, this is easily visible in E1 (Logarček), where conductivity in some puddles can reach more than 2000 μS/cm. This happens in winter, and the most probable source is the salty solution from the nearby motorway. A notable increase in SEC is also recorded in E1 (Vetrovna jama pri Laški kukavi) and W3 (Gašpinova jama), where pollution from the settlements above is possible (Fig. 5.62 and Table 5.4).

As with temperature, the conductivity recorded at most of the cave stations represents the local situation of the stagnant surface water, and only during high flow is a reading of the conductivity of the groundwater flow recorded (Fig. 5.62).

At positions with stagnant water (E1—Logarček, W1—Najdena jama, W2—Gradišnica, W3—Gašpinova jama, H1—Veliko brezno v Grudnovi dolini), the post-event SEC stabilises to a relatively low level as compared to the springs (S1—Močilnik, S2—Retovje, S3—Bistra) (Fig. 5.62 and Table 5.4). Some possible reasons for such low values are given in the discussion on SEC in H1 (Veliko brezno v Grudnovi dolini) (Sect. 5.3.1). However, as mentioned there, additional measurements of percolating water and cave air CO_2 would be needed.

The Unica River represents the main recharge to most of the observed system. Its conductivity is determined by the conductivities of different contributors and their mixing ratios. Table 5.4 gives the minimum and maximum conductivities of

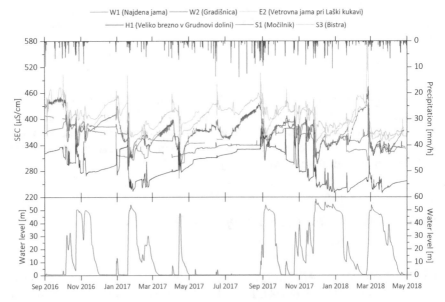

Fig. 5.62 Specific electrical conductivity (SEC) of water at selected observation points from September 2016 to May 2018. Missing values indicate periods when the water level was below the instrument. Lower plot shows the water level

Table 5.4 Measured minimal, maximal, and mean specific electrical conductivity (SEC) of water during selected periods

Name of observation point	Minimal SEC	Maximal SEC	Mean SEC	Mean SEC calculation period
Planinska jama—Pivka Branch	225	560	405	2016
Planinska jama—Rak Branch	315	520	415	2016
Planinska jama—spring of Unica River	255	520	370	2016
Malenščica River	324	464	370	2016
P2 (ponor Požiralnik 1/2 pod stenami)	295	430	377	Aug 2017–May 2018
W1 (Najdena jama)	265	410	340	2017
W2 (Gradišnica)	275	475	350	2017
W3 (Gašpinova jama)	250	790	365	2017
E1 (Logarček)	295	490	375	2016
E2 (Vetrovna jama pri Laški kukavi)	310	735	390	2017
H1 (Veliko brezno v Grudnovi dolini)	230	380	300	2017
S1 (springs Močilnik)	300	475	370	2017
S2 (springs Retovje)	320	465	390	Mar 2017–Mar 2018
S3 (springs Bistra)	355	475	410	2017

the tributaries Rak, Pivka, and Malenščica. SEC of the Unica River varies between 250 μS/cm during rain events and about 520 μS/cm during low water conditions (Fig. 5.63 and Table 5.4).

SEC increases along the flow between Planinsko Polje and the Ljubljanica River springs (Fig. 5.59) due to the dissolution of calcium carbonate along the flow pathway and due to mixing with autogenic recharge. Autogenic recharge mainly enters the system as a slow thin film flow, fracture flow or small vadose streams. It is generally well equilibrated with the limestone and therefore has high SEC. Above the cave Postojnska jama (southwest from cave Planinska jama), this is typically more than 400 μS/cm [13]. Finally, water springs out at the Ljubljanica River springs, where SEC varies from about 300 μS/cm during rain events to about 475 μS/cm during low water (Fig. 5.63 and Table 5.4).

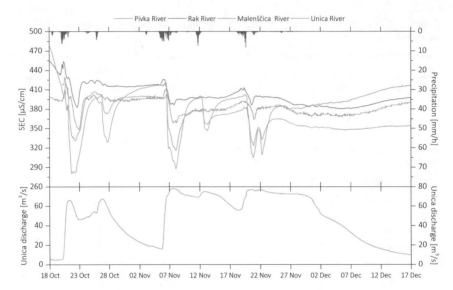

Fig. 5.63 An example of specific electrical conductivity (SEC) variations of the Unica River and its contributors during a high water event at the end of 2016

5.6.3   Temperature and Specific Electrical Conductivity Variations During High Water Events

High water events profoundly disturb the system. A typical event starts with the flushing out of previously stored water (equilibrated water) and continues with the arrival of relatively fresh, allogenic, and autogenic floodwater. A typical hydrograph of water temperature (T) and specific electrical conductivity (SEC) during a high water event in the studied aquifer consists of the following selected phases (Fig. 5.64):

1. **Small initial increase of water level; no response of T and SEC**. Onset of the rain event slightly increases the water level at all points. The response is purely hydraulic with no change in T or SEC (Fig. 5.64).
2. **Rapid increase of water level; initial small perturbation of T and small SEC peak**. Rapid recharge the aquifer causes inflow of the stored water and autogenic recharge with slightly different T and higher SEC (Fig. 5.64).
3. **Main flood stage; fast drop of SEC, which typically reaches a minimum and occurrence of diurnal oscillation of T**. Recharge by allogenic and autogenic inflow cause dilution and, therefore, a fast decrease in the SEC of the water (Fig. 5.64). Temperature of the water is typically irregular during the peak of flood events, but during recession (and weather clearance), the diurnal temperature signal from the surface is also transferred in the aquifer (Fig. 5.64). T and SEC of water in the caves follow the dynamics of the Unica River, where amplitudes are preserved, but delayed and damped by the distance from the ponors (Fig. 5.64).

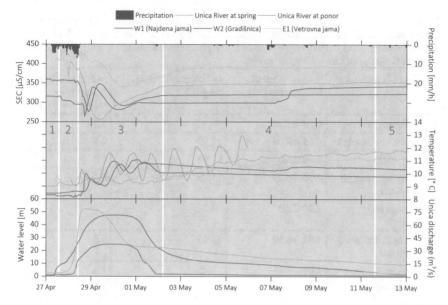

Fig. 5.64 Phases of water temperature (T) and specific electrical conductivity (SEC) dynamics during a high water event in spring 2017

4. **Recession of the water level; losing/maintaining the T and SEC signal of the Unica River**. If the active flow past the observation point is terminated, at that time, T and SEC signals of the Unica River are lost (Fig. 5.64). In the case of active underground flow, T and SEC signals are preserved, but damped and delayed (Fig. 5.64).

5. **Low water level; equilibrations to the local ambient conditions**. In caves with stagnant water, both T and SEC equilibrate to the local ambient conditions (Fig. 5.64).

5.6.4 Seasonal and Diurnal Temperature Variations

The temperature variations in the system are a consequence of the changing hydrological and climate conditions on the surface. High water events start with a typical disturbance of the temperature signal in the caves due to the mixing and arrival of different components. In the case of concentrated recharge from the surface, the initial disturbance is followed by a diurnal temperature signal, which can be traced deep into the system. These variations are appended to the annual temperature cycle, caused by seasonal changes of the climate.

Annual temperature variations are the largest in surface streams. At the official Hasberg gauging station, the Unica River has an annual variation of about 10 °C (Table 5.5), whereas at the ponors, just before sinking, the variation can reach up

Table 5.5 Annual and diurnal temperature variations of water in the observed ponors, caves, and springs in years 2015 to 2018. Larger variation and diurnal oscillation in springs Močilnik due to heat exchange, and smaller range of annual temperature variations in springs Retovje due to interrupted measurements in summer season

Name of observation point	Annual range of temperature [°C]	Difference in annual range [°C]	Maximal diurnal temperature oscillation [°C]
Unica River (Hasberg)	4.20–13.70	9.50	1.5
P1 (ponor Velike loke)	4.15–19.05	14.90	4.5
E1 (Logarček)	4.40–16.90	12.50	1.2
E2 (Vetrovna jama pri Laški kukavi)	4.50–16.25	11.75	0.9
E3 (Brezno pod Lipovcem)	8.20–8.50	0.30	/
P2 (ponor Požiralnik 2 pod stenami)	3.05–16.50	13.45	3.6
W1 (Najdena jama)	4.00–13.45	9.45	1.4
W2 (Gradišnica)	4.20–13.50	9.30	1.2
W3 (Gašpinova jama)	4.10–12.80	8.70	1.0
H1 (Veliko brezno v Grudnovi) dolini	7.70–9.25	1.55	/
H2 (Andrejevo brezno 1)	5.75–9.45	3.70	/
S1 (springs Močilnik)	5.95–13.85	7.90	1.4
S2 (springs Retovje)	5.70–11.95	6.25	/
S3 (springs Bistra)	7.00–13.60	6.60	/

to 15 °C (Table 5.5). At the ponors, the annual variations might be even larger, but during low water often drops below the position of instrument, so that temperatures cannot be observed. In caves with periodically active water flow, annual variations are diminishing by the distance from the ponors. They are from 8.7 °C in W3 (Gašpinova jama) to 12.5 °C in E1 (Logarček) (Table 5.5). In caves with stagnant water, the annual temperature variations are very small, from 0.3 °C in E3 (Brezno pod Lipovcem) to 3.7 °C in H2 (Andrejevo brezno 1) (Table 5.5). At the springs of the Ljubljanica River, annual variations are typically about 7 °C (Table 5.5).

In surface streams, diurnal temperature oscillations are present during the whole year, but the largest are during the summer months, when solar radiation is the strongest and discharge is the lowest (Fig. 5.65). At the Hasberg gauging station, the Unica River oscillates up to 1.5 °C (Table 5.5), but after a longer flow to the ponors, diurnal oscillation can be up to 4.5 °C (Table 5.5). In the studied water caves, diurnal temperature oscillations were mostly registered only during the recession stage of high water events (Fig. 5.64). In comparison with the ponors, they are delayed and damped. In the eastern regional groundwater flow, diurnal oscillation is diminished

Fig. 5.65 Slow meandering flow of the Unica River on Planinsko Polje. In spite of relatively abundant springs, the Unica River is prone to heat exchange due to a low gradient and slow meandering flow with a long total distance (17 km). Additionally, active ponors cause a gradual decrease in the amount of water (*Photo* M. Blatnik)

from 1.2 °C in E1 (Logarček) to 0.9 °C in E2 (Vetrovna jama pri Laški kukavi), while in the western regional flow, it is diminished from 1.4 °C in W1 (Najdena jama) to 1 °C in W3 (Gašpinova jama) (Table 5.5). In caves with stagnant water (E3—Brezno pod Lipovcem, H1—Veliko brezno v Grudnovi dolini, H2—Andrejevo brezno 1) and the springs of the Ljubljanica River, diurnal temperature oscillations were not detected (Table 5.5).

5.6.5   Transit Times and Water Flow Velocities

Temperature and SEC measurements can be used as a natural tracer. During periods of high water, sudden changes and/or diurnal variations of these parameters occur and are transferred along the surface and underground flow. Time differences in the transferred signals, and distances between the selected observation points, enable calculations of transit times and apparent groundwater flow velocities.

Transit times and apparent groundwater flow velocities are calculated for sectors between the studied ponors, water caves, and springs, which are presumably connected with groundwater flow (Fig. 5.66). Previous studies (for example, Gospodarič and Habič 1976) have identified two main regional water flows between

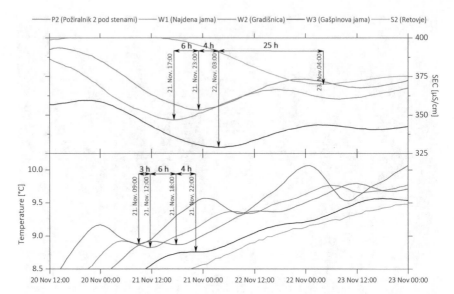

Fig. 5.66 Time differences in peaks of temperature (T) and specific electrical conductivity (SEC) variations, which enables calculations of transit times

Planinsko Polje and the Ljubljanica River springs. The eastern water flow connects the eastern ponors of Planinsko Polje (P1—Velike Loke) with caves E1 (Logarček), E2 (Vetrovna jama pri Laški kukavi), and the eastern springs of the Ljubljanica River (S3—Bistra) (Table 5.6 and Fig. 5.67). The western water flow connects the northern ponors of Planinsko Polje (P2—Pod stenami and P3—Škofov lom) with caves W1 (Najdena jama), W2 (Gradišnica), W3 (Gašpinova jama), and the western springs of the Ljubljanica River (S1—Močilnik and S2—Retovje) (Table 5.6 and Fig. 5.67).

Measurements of diurnal temperature oscillations were used for calculating groundwater flow between ponors and water caves, whereas changes in SEC were used for calculating further groundwater flow towards the springs of the Ljubljanica River, where diurnal temperature oscillations do not occur (Fig. 5.66). The actual distance of the groundwater flow between the observed points is not known, so the straight-line distance between observed points was used to assess the apparent flow velocity (Table 5.6 and Fig. 5.67).

Transit Times Between Consecutive Points

The straight-line distance of the eastern regional groundwater flow is about 11.2 km, and the time difference between evident changes of the SEC from the eastern ponors of Planinsko Polje (P1) to the springs of the Ljubljanica River is from 52 h at S2 (springs Retovje) to 61 h at E3 (springs Bistra) (Table 5.6 and Fig. 5.67). The calculated apparent groundwater flow velocity is, therefore, from 180 to 215 m/h. The calculations from the temperature oscillations show that the first part of the groundwater flow is relatively slow. The flow velocity between the eastern ponors on Planinsko Polje (P1) and E1 (cave Logarček) is about 140 m/h (Table 5.6 and

Table 5.6 Straight-line distances, water level drops, transit times, and apparent water flow velocities between consecutive observation points during high water periods

Sector		Straight-line distance (m)	Drop of water level during high water level event (m)	Transit time (h)	Apparent flow velocity (m/h)
Eastern regional flow (sectors)	P1–E1 (eastern ponors of Planinsko Polje–Logarček)	1500	3 (0.002)	10.0–11.5	130–150
	E1–E2 (Logarček–Vetrovna jama pri Laški kukavi)	1600	14 (0.009)	7.7–8.4	190–210
	E2–S2/S3 (Vetrovna jama pri Laški kukavi–eastern springs of the Ljubljanica River)	8100	132 (0.016)	34–43	190–240
	P1–S2/S3 (eastern ponors of Planinsko Polje –eastern springs of the Ljubljanica River)	**11,200**	**149 (0.013)**	**52–61**	**185–215**
Western regional flow (sectors)	P2/P3–W1 (northern ponors of the Planinsko Polje–Najdena jama)	450	12 (0.027)	4.2–4.4	100–110
	W1–W2 (Najdena jama–Gradišnica)	2000	3 (0.002)	6.2–6.4	310–325
	W2–W3 (Gradišnica–Gašpinova jama)	1600	7 (0.004)	5.6–6.1	260–285
	W3–S1/S2 (Gašpinova jama–western springs of the Ljubljanica River)	7050	127 (0.018)	22.5–29.5	240–315
	P2/P3–S1/S2 (northern ponors of Planinsko Polje –western springs of the Ljubljanica River)	**11,100**	**149 (0.013)**	**39–46**	**240–285**

Bold text present total values for separate sections, which are not written with bold text

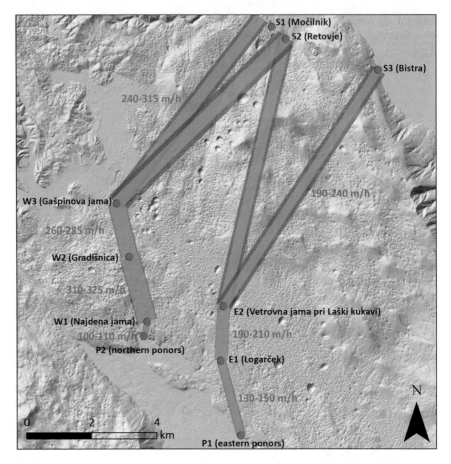

Fig. 5.67 Ranges of apparent groundwater flow velocities between selected observation points during various high water events (DEM data from ARSO [2]; Cave data from Cave Register [5])

Fig. 5.67). Further flow from E1 (Logarček) towards E2 (Vetrovna jama pri Laški kukavi) and the springs of the Ljubljanica River (S2 and S3) is notably faster—from 180 to 240 m (Table 5.6 and Fig. 5.67).

The straight-line distance of the western regional groundwater flow is about 11.1 km, and the average time difference between evident peaks of the SEC from the western ponors on Planinsko Polje (P2 and P3) to the springs of the Ljubljanica River is from 39 h at S2 (springs Retovje) to 46 h at S1 (springs Močilnik) (Table 5.6 and Fig. 5.67). The calculated apparent groundwater flow velocity is, therefore, from 240 to 285 m/h. Like with the eastern regional groundwater flow, the apparent flow velocity between ponors (P2 and P3) and W1 (Najdena jama) is relatively small— about 100 m/h, but the drop of the water level during high water events is relatively large—12 m at 450 m of straight-line distance (Table 5.6 and Fig. 5.67). The flow between caves W1 (Najdena jama), W2 (Gradišnica), and W3 (Gašpinova jama) is much faster (between 260 and 315 m/h) (Table 5.6 and Fig. 5.67), which indicates well developed channels. The drop of the water level between W3 (Gašpinova jama) and the western springs of the Ljubljanica River (S1 and S2) during high water events is high (127 m of drop at 7050 m of straight-line distance), and some characteristics of the water level measurements show that there is probably at least one hydrogeological barrier. In this section, apparent groundwater flow velocity is high and varies vary from 240 to 315 m/h (Table 5.6 and Fig. 5.67).

Groundwater flow velocity evidently depends on the discharge (Fig. 5.68). The highest water flow velocities are present when Planinsko Polje is flooded and the water level in the aquifer is the highest. During recession, the flow velocity slows down until the limit of assessment is reached. For example, the shortest transit time of the water flow between the northern ponors (P2 and P3) and W2 (cave Gradišnica) was about 10.5 h, while the largest was about 30 h. The shortest transit time of the water flow between the eastern ponors (E1) and E2 (cave Vetrovna jama pri Laški kukavi) was about 16 h, while the largest was about 35 h (Fig. 5.68). Figure 5.68 presents a set of transit time estimations during a high water event that occurred in March and April 2015. When the polje was flooded, transit times of the groundwater flow were stable. When the eastern group of ponors were switched-off (11–13 April), transit times immediately and evidently increased (Fig. 5.68).

Transit Times Between Ponors and Springs
Analysis of high water events showed that transit times from P2 (northern ponors of Planinsko Polje) to S2 (springs Retovje) varied from 32 to 55 h (on average 39 h). This presents a relatively large difference in apparent water flow velocities (200–350 m/h, on average 280 m/h) (Table 5.7). Transit times from P2 (northern ponors of Planinsko Polje) to S3 (springs Močilnik) were longer and varied from 40 to 62 h (on average 46 h). Calculated apparent water flow velocities had a range from 180 to 280 m/h (on average 240 m/h) (Table 5.7). The large range of apparent water flow velocities is strongly related to the hydrological situation. The measurements indicate that a high discharge of the Ljubljanica River is related to faster groundwater flow (Fig. 5.69).

Transit times between P1 (eastern ponors of Planinsko Polje) and S2 (springs Retovje) varied from 44 to 69 h (on average 52 h) with the apparent flow velocities ranging from 160 to 255 m/h (on average 215 m/h) (Table 5.7). The time difference between P1 (eastern ponors of Planinsko Polje) and S3 (springs Bistra) ranged between 44 and 82 h (on average 61 h), giving groundwater flow velocities from 140 to 255 m/h (on average 185 m/h) (Table 5.7).

Transit Times in Comparison to Previous Measurements
The observed velocities can be, to some extent, compared to the values obtained during the dye tracing campaign in the 1970s (Gospodarič and Habič 1976). During that time, the dominant flow velocity between P2 (Pod stenami) and S2 (spring Pod skalo—Retovje) was 166 m/h and the maximum velocity was 191 m/h (Table 5.8 and

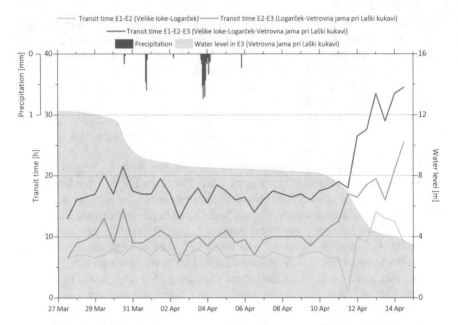

Fig. 5.68 Transit times of the groundwater flow between ponor P1 (Velike loke) and caves E1 (Logarček) and E2 (Vetrovna jama pri Laški kukavi) during a flood event in March and April 2015, estimated from the phase shift of diurnal temperature records. Blue background represents the water level at E2 (Vetrovna jama pri Laški kukavi)

Table 5.7 Range of transit times and apparent groundwater flow velocities between the studied ponors and springs

Sector	Straight-line distance (m)	Transit time (h)	Apparent flow velocity (m/h)
P1–S2 (eastern ponors of Planinsko Polje–springs Retovje)	11,200	44–69 (52)	160–255 (215)
P1–S3 (eastern ponors of Planinsko Polje–springs Bistra)	11,200	44–82 (61)	140–255 (185)
P2/P3–S1 (northern ponors of Planinsko Polje–springs Močilnik)	11,100	40–62 (46)	180–280 (240)

Numbers in brackets show average value

Fig. 5.70). This is evidently lower than the above presented results (200–350 m/s) (Table 5.8). Similar velocities were registered in the sector between P2 (Pod stenami) and S1 (Močilnik). During the dye tracing campaign, the peak velocity was 162 m/h, while the maximum velocity was 191 m/h (Table 5.8 and Fig. 5.70). These values are also relatively low compared to the new measurements (190–280 m/s) (Table 5.8). In

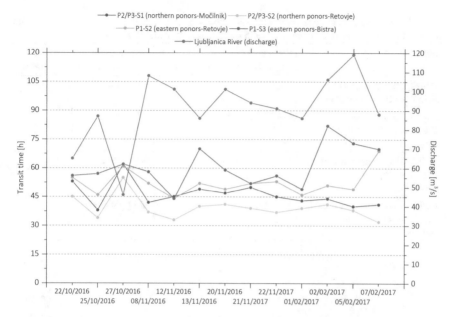

Fig. 5.69 Different durations of groundwater transit times between selected ponors and springs at various hydrological conditions

the sector between ponors Dolenje Loke (comparable with P1—Velike Loke in the new study) and S1 (spring Pod skalo—Retovje), the peak velocity of the tracer was 194 m/h, while the maximum velocity was 299 m/h (Table 5.8 and Fig. 5.70). There, the results returned higher values than the new values obtained with continuous measurements (160–255 m/s) (Table 5.8). The most similar results between the two studies were calculated for the sector between ponor Dolenje loke (comparable with P1—Velike Loke in the new study) and E1 (cave Logarček). The peak velocity of the tracer in 1975 was 144 m/h (Table 5.8 and Fig. 5.70), while new measurements show water flow velocities between 130 and 150 m/h (Table 5.8).

The values obtained in this study are systematically larger, as the method allows estimation only at relatively high water when active flow at the observation points is present. For most caves, discharge of the Unica River needs to be at least 40 m³/s. Moreover, many high water events were included in the analysis. At the time of tracing, one high water event was analysed (27 May 1975) and the discharge of the Unica River was about 20 m³/s (ARSO 2018a). Additionally, the propagation and transmission of signals of dye tracers, temperature and specific electrical conductivity (SEC) differ and can result in different transit times (Bekele et al. 2014).

Table 5.8 Comparison of apparent groundwater flow velocities obtained in the present study with the water tracing experiment from May 1975 (after Gospodarič and Habič 1976)

Sector	SUWT 1975 Apparent flow velocity (m/h)	Present study Apparent flow velocity (m/h)
P1–E1 (eastern ponors of Planinsko Polje–Logarček)	144	140 (130–150)
P1–S2 (eastern ponors of Planinsko Polje–springs Retovje)	194 (max 299)	185 (140–255)
P2/P3–S1 (northern ponors of Planinsko Polje–springs Močilnik)	162 (max 191)	240 (180–280)
P2/P3–S2 (northern ponors of Planinsko Polje–springs Retovje)	166 (max 191)	285 (200–350)

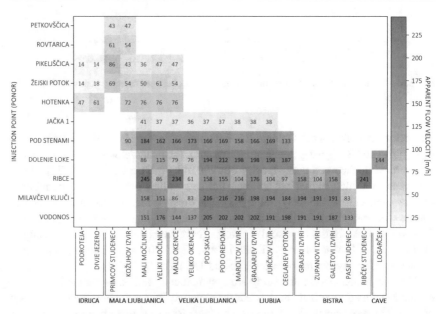

Fig. 5.70 Apparent flow velocities between the studied ponors and the springs obtained from the dye tracing campaign in 1975 (adapted from Gospodarič and Habič 1976). On vertical axis injection points (ponors or surface streams) whereas on horizontal axis locations of sampling points (springs or caves) are listed. Filled squares with numbers indicate proven connections and dominant flow velocities (expressed in m/h)

5.7   Groundwater Flow Directions in the Area—A Brief Review of Past and New Findings

The aquifer between Planinsko Polje and the springs of the Ljubljanica River has many inflows and outflows from the system. Past and recent studies of this aquifer indicate convergence, bifurcation, and interweaving of several groundwater pathways. The results of the tracing campaign in the 1970s (Gospodarič and Habič 1976) (Fig. 5.71a) can be summarised as the following:

- Flow from Cerkniško Polje is joined by part of the flow from the eastern ponors of Planinsko Polje. These inputs are the main feeders of the Bistra spring. Their relative contribution to other springs diminishes in an east–west direction.
- Flow from the northern ponors of Planinsko Polje contributes most to the western spring group, while almost no contribution to Bistra was detected.
- The western group of springs also receives water from Logaško Polje, Rovte Plateau, and Hotenka Stream.

There is a clear distinction between the eastern and western flow pathways, although the results indicate convergence and bifurcation of these flows (Fig. 5.71a).

In 2006 and 2007, water level and temperature observations were acquired in four caves in the studied aquifer [21]. The results indicated possible directions of the groundwater flow and hydraulic connections between the caves selected for

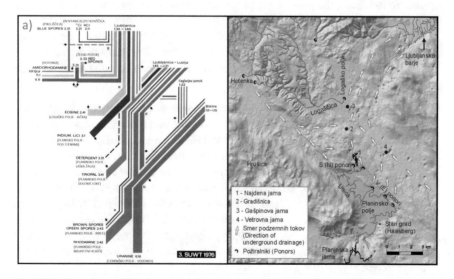

Fig. 5.71 Groundwater directions according to previous studies: **a** the groundwater connections scheme according to the results of water tracing tests in 1975 (from Gospodarič and Habič 1976, Plate XI); **b** groundwater pathways based on the water level and temperature measurements in the caves in 2006 and 2007 (from Turk [21]: 108)

study. Three possible directions of the groundwater flow were identified [10, 19, 21] (Fig. 5.71b):

- During high water, a clear connection between the northern ponors of the Planinsko Polje, Najdena jama, and Gradišnica is established. Gradišnica and Gašpinova jama are hydraulically well-connected. The uniform changes of the level in these two caves indicate a possible flow barrier downstream from Gašpinova jama.
- There is a connection between the eastern ponors of Planinsko Polje and Vetrovna jama pri Laški kukavi. The maximal level in Vetrovna jama pri Laški kukavi was related to the unknown underground overflow or to the limited recharge from Planinsko Polje.
- Turk [21] also mentions a possible groundwater connection between the eastern ponors of Planinsko Polje and the cave Gradišnica.

Compared to the previous studies, the present study comprises a larger number of water caves (8), ponors (4), and springs (3), in addition to a longer observation period (3.5 years) and more measured parameters (specific electrical conductivity, as well as water level and temperature). These improvements along with the favourable hydrological conditions, with periods of very low water level alternated by more prominent high water events, provided a large amount of useful data. Further analysis and additional modelling of the data provide new findings including the identification of potentially new directions of groundwater flow, hydraulic connections between a number of the caves, the presence of potential hydrological barriers, and differences during various hydrological conditions.

Most of the findings, already presented in the previous sections, are summarized in an improved map of possible groundwater flow directions and connections (Fig. 5.72). New findings, including possible directions of the groundwater flow, are summarized in the following paragraphs:

- **Groundwater flow related to eastern ponors of the Planinsko Polje**. Flow from Planinsko Polje enters the aquifer through a broad outflow region (E1) and continues towards caves E1 (Logarček) and E2 (Vetrovna jama pri Laški kukavi) (Fig. 5.72). During low flow, the connection between both caves is not clear. During high flow, all water level hydrographs show a clear connection between the two caves along the base channel and an important overflow level. The presence of the latter gives rise to a specific relation between the water level hydrographs in both caves. During flood events, the water level in E1 (Logarček) almost reaches the level of Planinsko Polje. This clearly shows that discharge from the polje is limited by the transmissivity of the aquifer and not by the local capacity of the ponor zones.

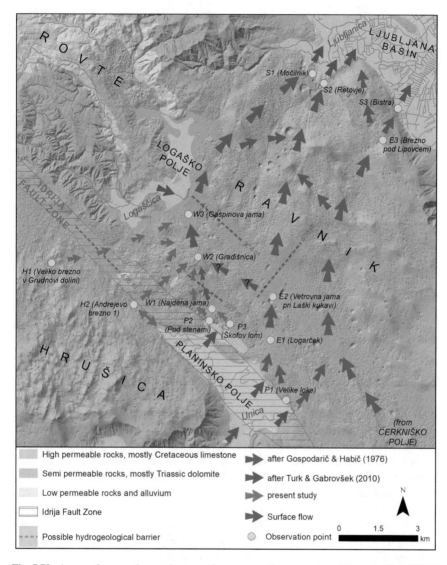

Fig. 5.72 A map of assumed groundwater pathways according to previous (Gospodarič and Habič 1976) [21] and new measurements (Geological data from Krivic et al. [16]; DEM data from ARSO [2]; Cave data from Cave Register [5])

- **Groundwater flow related to the northern ponors of Planinsko Polje**. Flow from ponor zone P2 (Pod stenami) bypasses the cave W1 (Najdena jama) and flows directly towards the block with W2 (Gradišnica) and W3 (Gašpinova jama) (Fig. 5.72). When the outflow at P2 (Pod stenami) is surpassed by the recharge, the flow is diverted along the higher positioned channels towards the ponor zone P3 (Škofov lom), which has a clear connection to W1 (Najdena jama) (Fig. 5.72).

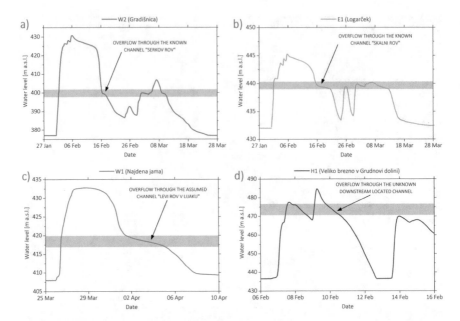

Fig. 5.73 Inflections of water level hydrographs allowed to identify the overflow phenomena through the known (**a** and **b**), assumed (**c**) and potential yet unknown (**d**) passages

Response in the block with W2 (Gradišnica) and W3 (Gašpinova jama) precedes the response of ponor zone P2 (Pod stenami). This initial response might be related to the connection to the eastern flow from the block with E1 (Logarček) and E2 (Vetrovna jama pri Laški kukavi) or to the recharge from the Hrušica Plateau, detected in H1 (Veliko brezno v Grudnovi dolini) (Fig. 5.72). New observations confirm the assumptions of [21] and Gabrovšek (2010) that caves W2 (Gradišnica) and W3 (Gašpinova jama) belong to a block with uniform water level.

- **Groundwater flow related to the Hrušica Plateau.** The water level hydrograph from H1 (Veliko brezno v Grudnovi dolini) indicates a flow barrier with a highly transmissive zone above it. The flow from the Hrušica Plateau is easily transferred along this zone towards the block with caves W2 (Gradišnica) and W3 (Gašpinova jama) (Fig. 5.72). It is the flooding in this zone that causes a vigorous response in H1 (Veliko brezno v Grudnovi dolini) and also in H2 (Andrejevo brezno 1). The highest back floods surpass the level of Planinsko Polje and give rise to numerous springs below the village of Grčarevec (Fig. 5.72). The recession in the block, related to floods in H2 (Andrejevo brezno 1), precedes the recession in Planinsko Polje, which results in outflow from Planinsko Polje towards H2 (Andrejevo brezno 1) (Fig. 5.72).

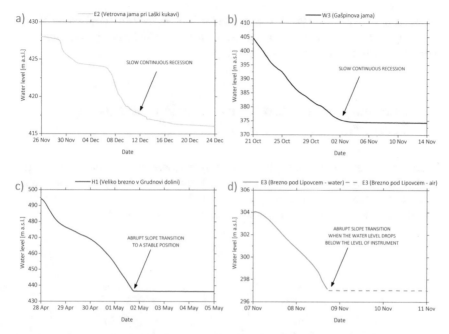

Fig. 5.74 Recession of water level hydrographs is a complex phenomenon driven by the input–output relations. The terminate part typically show exponential decrease towards the stable position (**a** and **b**). In some cases, the slope of the recession curve abruptly flattens, which can be a result of different mechanisms: in **c** the water drops to the position of a flow barrier which keeps the local water table extremely stable, whilst in **d** the water level drops below the level of instrument

- **Possible groundwater flow between W2 (Gradišnica) and E2 (Vetrovna jama pri Laški kukavi)**. Previous studies indicated possible groundwater flow between these caves (Fig. 5.72). In spite of an almost simultaneous response in water level increase (sooner than in the northern ponors of Planinsko Polje), recent observations cannot reliably confirm active flow between these two locations.

One of the objectives of this research was to correlate the observations to the geological structure of the area. However, with some rare exceptions, this turned out to be a task beyond the scope of this work, as extremely detailed geological mapping is required to establish the connection between the present-day geometry of the flow system to the geological structure.

The uniform and stable water levels present in some caves indicate flow barriers, which are most probably conditioned by the geological structure (Fig. 5.72). In the case of H1 (Veliko brezno v Grudnovi dolini), with a stable base level at 436 m a.s.l., the flow barrier may be the Idrija Fault Zone, which obviously has at least a locally very transmissive zone above this level (Fig. 5.72). The very stable level in W2 (Gradišnica) and W3 (Gašpinova jama) also indicate such a barrier (Fig. 5.72), although it is hard to relate it to a particular lithological/structural element. The relatively uniform water level in most sumps in W1 (Najdena jama) also indicates a barrier downstream from this cave [17] (Fig. 5.72).

5.8   Summary of the Most Characteristic Hydrograph Features

Long-term observations allow determining hydrograph feature, which are characteristic for all events of certain intensity and duration and may be related to the geometric feature of the aquifer. Figures 5.73, 5.74, 5.75 and 5.76 present a small selection of characteristic level and temperature hydrographs.

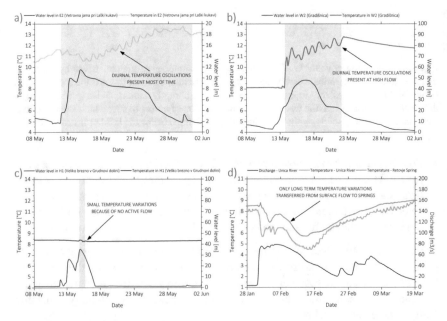

Fig. 5.75 Some examples of temperature hydrographs: **a** when water from polje (or allogenic stream) continuously flow by the instrument the diurnal cycle may be present through most of the record; **b** diurnal oscillations occur only at high flow when active flow at the instrument is established through overflow passages; **c** in some cases only small variations of temperature were detected; there is no concentrated input from the surface and/or there is no active flow past the instrument; **d** at the springs diurnal and other short term signals are not visible. Only long-term changes are transferred that far

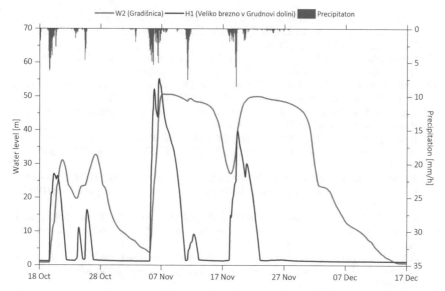

Fig. 5.76 Water level hydrographs of W2 (Gradišnica) and H1 (Veliko brezno v Grudnovi dolini) during a series of rain events in autumn 2016. The hydrographs show the role of Planinsko Polje, which acts as a natural retention basin. The prolonged outflow from the flooded polje keeps the level in W2 high during the events. In contrast, the level in H1 drops fast due to the efficient drainage and lack of large storage elements

References

1. ARSO (2018) Meteorological archive [Online]. Available from: https://meteo.arso.gov.si/met/sl/archive/. Accessed 7 Apr 2020
2. ARSO (2018) Lidar data fishnet [Online]. Available from: https://gis.arso.gov.si/. Accessed 7 Apr 2020.
3. Badino G (2010) Underground meteorology—what's the weather underground? Acta Carsologica 39(3):427–448. https://doi.org/10.3986/ac.v39i3.74
4. Blatnik M, Frantar P, Kosec D, Gabrovšek F (2017) Measurements of the outflow along the eastern border of Planinsko Polje, Slovenia. Acta Carsologica 46(1):83–93. https://doi.org/10.3986/ac.v46i1.4774
5. Cave Register (2018) Cave register of the Karst Research Institute ZRC SAZU and Speleological Association of Slovenia. Postojna, Ljubljana
6. Covington MD, Perne M (2015) Consider a cylindrical cave: a physicist's view of cave and karst science. Acta Carsologica 44(3):363–380. https://doi.org/10.3986/ac.v44i3.1925
7. Covington MD, Luhmann AJ, Gabrovšek F, Saar MO, Wicks CM (2011) Mechanisms of heat exchange between water and rock in karst conduits. Water Resour Res 47:W10514. https://doi.org/10.1029/2011WR010683
8. Covington MD, Luhmann AJ, Wicks CM, Saar MO (2012) Process length scales and longitudinal damping in karst conduits. J Geophys Res 117:F01025. https://doi.org/10.1029/2011JF002212
9. Ford DC, Williams PW (2007) Karst hydrogeology and geomorphology. Willey, Chichester, p 562

10. Gabrovšek F, Turk J (2010) Observations of stage and temperature dynamics in the epiphreatic caves within the catchment area of the Ljubljanica River (Slovenia). Geol Croat 63(2):187–193. https://doi.org/10.4154/gc.2010.16

11. Gabrovšek F, Peric B, Kaufmann G (2018) Hydraulics of epiphreatic flow of a karst aquifer. J Hydrol 560:56–74. https://doi.org/10.1016/j.jhydrol.2018.03.019

12. Gospodarič R, Habič P (eds) (1976) Underground water tracing: investigations in Slovenia 1972–1975. Inštitut za raziskovanje krasa ZRC SAZU, Postojna, p 312

13. Kogovšek J (2010) Characteristics of percolation through the karst vadose zone. Inštitut za raziskovanje krasa ZRC SAZU, Postojna, p 168

14. Krawczyk WE, Ford DC (2006) Correlating specific conductivity with total hardness in limestone and dolomite karst waters. Earth Surf Proc Land 31:221–234. https://doi.org/10.1002/esp.1232

15. Krajnc B, Ferlan M, Ogrinc N (2016) Soil CO_2 sources above a subterranean cave—Pisani rov (Postojna Cave, Slovenia). J Soils Sediments 17(7):1883–1892. https://doi.org/10.1007/s11368-016-1543-x

16. Krivic P, Verbovšek R, Drobne F (1976) Hidrogeološka karta 1:50000. In: Gospodarič R, Habič P (eds) Underground water tracing: investigations in Slovenia 1972–1975. Inštitut za raziskovanje krasa ZRC SAZU, Postojna

17. Šušteršič F (1981) Morfologija in hidrografija Najdene jame. Acta Carsologica 10:127–156

18. Šušteršič F (2002) Where does underground Ljubljanica flow? RMZ Mater Geoenviron 49(1):61–84

19. Turk J (2008) Hidrogeologija Gradišnice in Gašpinove jame v kraškem vodonosniku med Planinskim poljem in izviri Ljubljanice. Geologija 51(1):51–58

20. Turk J (2010a) Hydrogeological role of large conduits in karst drainage system. University of Nova Gorica, Graduate School, Nova Gorica, p 305

21. Turk J (2010b) Dinamika podzemne vode v kraškem zaledju izvirov Ljubljanice - Dynamics of underground water in the karst catchment area of the Ljubljanica springs. Inštitut za raziskovanje krasa ZRC SAZU, Postojna, p 136

Chapter 6
Conclusions

A network of autonomous measurements of groundwater parameters in the Ljubljanica River recharge area was set up. The observations were established and maintained for a period of three years in four ponors, eight active caves, and three springs. The records of water level, temperature, and specific electrical conductivity enabled new insights into the structure and mechanisms of this complex aquifer.

Data interpretation and hydraulic modelling led to new conclusions on the structure of the observed system, flow paths, and flood dynamics. Some of these are summarized in the following paragraphs.

- **Importance of overflow channels:**

  - During floods, higher positioned epiphreatic channels are activated. Such overflows are recorded as the inflection in the water level hydrographs at the elevation of the overflow channel. In several caves, the water level at the curve inflection matches the position of known overflow passages. Other inflection levels indicate possible positions of yet unknown passages (see Sects. 5.1, 5.2, and 5.3).
  - The water surface between Planinsko Polje and the town of Logatec changes from concave to convex during floods. This may point to the role of the overflow channels, which seem to be more common in the upper part of the system and probably scarcer in the lower part (see Sects. 5.1 and 5.2).
  - Known and unknown overflow levels were key for the interpretation when the correlating water levels in the caves is aligned along the same flow paths (see Sects. 5.1 and 5.2).

- **Distribution of flow from Planinsko Polje into the aquifer:**

  - Using a novel approach, the outflow from Planinsko Polje into the eastern set of ponors was estimated.
  - In contrast to the expectations, the results indicate direct connection between ponor zone P2 (Pod Stenami) and W2 (Gradišnica). Only when this pathway

© The Editor(s) (if applicable) and The Author(s), under exclusive license   151
to Springer Nature Switzerland AG 2020
M. Blatnik, *Groundwater Distribution in the Recharge Area*
of Ljubljanica Springs, Springer Theses, https://doi.org/10.1007/978-3-030-48336-4_6

is surcharged is the recharge to nearby W1 (Najdena jama) activated through ponor zone P3 (Škofov lom), due to the diversion of flow to the higher positioned channels in Planinsko Polje (see Sect. 5.2.1).

- The hydraulic gradient along the eastern border of Planinsko Polje is very low during high floods. The level in E1 (Logarček) is only a few meters below the polje (see Sect. 5.1.1).

- **The flow from the Hrušica Plateau and the role of Idrija Fault Zone:**

  - Water level in H1 (Veliko brezno v Grudnovi dolini) has a very stable position during low flow with vigorous response at the events and a rapid recession, ending with an abrupt transition to the base level (see Sect. 5.3).
  - The behaviour was explained with a flow barrier below a highly transmissive level. The fast response in H1 (Veliko brezno v Grudnovi dolini) occurs when the water on the northern side rises above the barrier and back-floods the area of the observed cave. A hydraulic model was developed which successfully captured this behaviour, as well as the interaction between Planinsko Polje and the flow from the Hrušica Plateau, which is mediated through the set of estavelles below the village of Grčarevec (see Sect. 5.3).

- **Use of temperature and specific electrical conductivity (SEC) hydrographs as natural tracers:**

  - Based on the temperature and SEC hydrographs, groundwater velocities between different observation points were estimated. Although, the estimation is only possible for the period of high water conditions. Calculated apparent velocities mostly ranked between 200 and 300 m/h. Temperature signals were also used to detect overflowing phenomena (see Sect. 5.6).
  - Temperature hydrographs of the springs also confirmed some of the previously determined flow connections, such as the dominant connection between Cerkniško Polje and the springs of Bistra (see Sect. 5.4).

This work demonstrates the potential of active water caves as an integral part of groundwater monitoring in karst. Combining information from the level, temperature, and specific electrical conductivity hydrographs, existing knowledge on the geometry of conduit systems, and (relatively simple) numerical models, gave new insights into the local and regional structure of the karst aquifer, the dynamics of flood events, and a much more detailed picture of flow paths. These findings enable better assessment of the water balance, water quality, and better understanding of the flooding of Planinsko Polje. The methodology developed in this work could be applied to any karst aquifer with dominant conduit flow.

However, there is still a lot of room for the improvement of the methodology:

- The local characteristics of observation points play an important role. Often it is not possible to get a good temperature signal in all hydrological situations and high sensitivity to level changes at the same spot. Therefore, careful consideration is needed prior to setting up the observation point. As the instruments are becoming cheaper and new data transfer technologies are becoming available, a higher number of instruments could provide better insight in different hydrological situations.

- Hydrograph analyses in this work were mainly qualitative, with limited and basic use of time series analyses. Advanced data modelling would surely provide new results, which were overseen in this work. These possibilities are still to be explored.
- The numerical modelling used in this work was quite basic, although good qualitative and in some cases also quantitative fits were obtained. The methods used were based on the assumption of conduit dominated flow in the epiphreatic zone, which is acceptable during high flow. One must be aware that this is just an approximation and that fracture and matrix flow need to be considered for a complete picture.
- More advanced and complex models could also reveal a finer structure of an aquifer and more detailed dynamics of groundwater flow.

Many open questions also remain within the system. Some can surely be answered based on the data obtained in this study. Advanced time series analyses and correlations with weather, geology, and structures could give new results. New speleological explorations might give access to new observation points. It has been shown here that each additional point resulted in new conclusions. For example, the western (Hrušica Plateau) block could only be understood after the observation in cave H2 (Andrejevo brezno 1) started. The role of W1 (Najdena jama) was unclear until observations in ponor zone P3 (Škofov lom) proved that its block is primarily recharged from there. New caves with access to the groundwater west of W1 (Najdena jama) and W2 (Gradišnica), and between W3 (Gašpinova jama) and the springs of the Ljubljanica River, could aid in constraining the present models. Furthermore, the open question regarding the nature of the connection between the eastern and western flow could be resolved by making observations in caves between E2 (Vetrovna jama pri Laški kukavi) and the block of W2 (Gradišnica) and W3 (Gašpinova jama).

Furthermore, it needs to be mentioned that the structure and extent of the phreatic zone and deep regional flow were not discussed in this work and remain a major challenge for the future.

Appendix A

Conceptual model and settings used for the hydraulic model of the system between Planinsko Polje and caves E1 (Logarček) and E2 (Vetrovna jama pri Laški kukavi). Some characteristics of nodes, used in conceptual and numerical model.

Name	Type of node	Invert elevation of node in m	Initial depth (initial water level above invert in m)	Constant (area of storage unit in m^2)
Node P1 (Planinsko Polje)	Storage unit with inflow	442	0	1000
Node E1 (Logarček)	Storage unit	432	0	1000
Node E2 (Vetrovna jama pri Laški kukavi)	Junction	415	0	/
Outflow	Outfall	405	0	/

Some characteristics of links, used in conceptual and numerical model.

Name	Maximal depth (diameter of conduit in m)	Length of conduit in m	Inlet offset (position of inlet of channel above the invert in m)	Outlet offset (position of outlet of channel above the invert in m)
C1 (P1–E1)	4	1500	0	0
C2 (E1–E2)	2.4	1500	0	0
C3 (E1–E2)	1.6	2500	7	6
C4 (E1–>outflow)	2.7	2000	0	0

M. Blatnik, *Groundwater Distribution in the Recharge Area of Ljubljanica Springs*, Springer Theses, https://doi.org/10.1007/978-3-030-48336-4

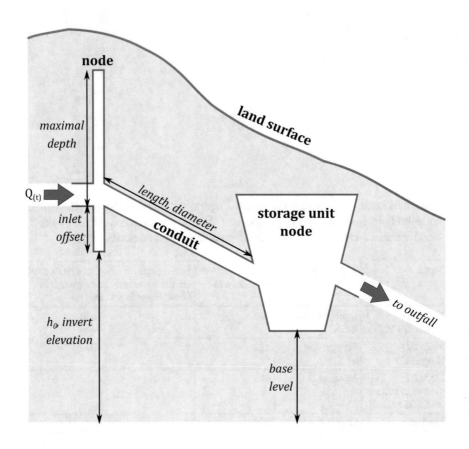

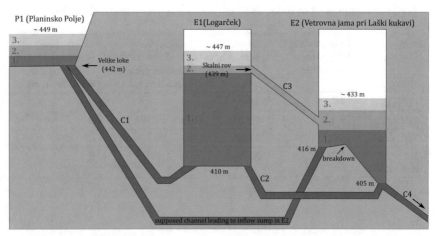

Appendix B

Conceptual model and settings used for the hydraulic model of the system between Planinsko Polje and caves W1 (Najdena jama) and W2 (Gradišnica).

Some characteristics of nodes, used in conceptual and numerical model.

Name	Type of node	Invert elevation of node in m	Initial depth (initial water level above invert in m)	Constant (area of storage unit in m^2)
Node P2/P3 (Planinsko Polje)	Storage unit with inflow	430	0	/
Node W1 (Najdena jama)	Storage unit	408	0	10,000
Node W2 (Gradišnica)	Storage unit	377	0	2000
Node W3 (Gašpinova jama)	Storage unit	374	0	50,000
Outflow	Outfall	370	0	/

Some characteristics of links, used in conceptual and numerical model.

Name	Maximal depth (diameter of conduit in m)	Length of conduit in m	Inlet offset (position of inlet of channel above the invert in m)	Outlet offset (position of outlet of channel above the invert in m)
C1 (P2/P3–W2)	3	2000	0	0
C2(P2/P3–W1)	10	400	0	0
C3 (W1–W2)	3.25	1500	0	0

(continued)

M. Blatnik, *Groundwater Distribution in the Recharge Area of Ljubljanica Springs*, Springer Theses, https://doi.org/10.1007/978-3-030-48336-4

(continued)

Name	Maximal depth (diameter of conduit in m)	Length of conduit in m	Inlet offset (position of inlet of channel above the invert in m)	Outlet offset (position of outlet of channel above the invert in m)
C4 (W1–W2)	4	1500	10	0
C5 (W2–W3)	4.25	2000	0	0
C6 (W2–W3)	4	2000	23	0
C7 (W3–outflow)	3.8	2000	0	0

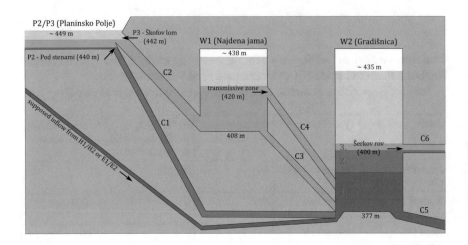

Appendix C

Conceptual model and settings used for the hydraulic model of the system between caves W2 (Gradišnica) and W3 (Gašpinova jama).

Some characteristics of nodes, used in conceptual and numerical model.

Name	Type of node	Invert elevation of node in m	Initial depth (initial water level above invert in m)	Constant (area of storage unit in m^2)
Node W2 (Gradišnica)	Junction with inflow	2	0	/
Node W3 (Gaspinova jama)	Junction	1	0	/
Outflow	Outfall	30	0	/

Some characteristics of links, used in conceptual and numerical model.

Name	Maximal depth (diameter of conduit in m)	Length of conduit in m	Inlet offset (position of inlet of channel above the invert in m)	Outlet offset (position of outlet of channel above the invert in m)
C1 (W2–W3)	1.5	400	0	0
C2 (W2–W3)	5	400	30	29
C3 (W3–outflow)	1.5	400	0	0

© The Editor(s) (if applicable) and The Author(s), under exclusive license
to Springer Nature Switzerland AG 2020
M. Blatnik, *Groundwater Distribution in the Recharge Area
of Ljubljanica Springs*, Springer Theses, https://doi.org/10.1007/978-3-030-48336-4

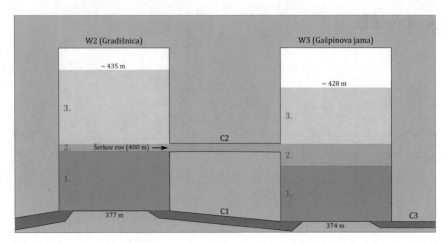

Appendix D

Conceptual model and settings used for the hydraulic model of the system between Planinsko Polje and caves W1 (Najdena jama), W2 (Gradišnica), and W3 (Gašpinova jama).

Some characteristics of nodes, used in conceptual and numerical model.

Name	Type of node	Invert elevation of node in m	Initial depth (initial water level above invert in m)	Constant (area of storage unit in m^2)
Node P2/P3 (Planinsko Polje)	Storage unit with inflow	430	0	/
Node W1 (Najdena jama)	Storage unit	408	0	10,000
Node W2 (Gradišnica)	Storage unit	377	0	2000
Node W3 (Gašpinova jama)	Storage unit	374	0	50,000
Outflow	Outfall	370	0	/

Some characteristics of links, used in conceptual and numerical model.

Name	Maximal depth (diameter of conduit in m)	Length of conduit in m	Inlet offset (position of inlet of channel above the invert in m)	Outlet offset (position of outlet of channel above the invert in m)
C1 (P2/P3–W2)	3	2000	0	0
C2 (P2/P3–W1)	10	400	0	0

(continued)

M. Blatnik, *Groundwater Distribution in the Recharge Area
of Ljubljanica Springs*, Springer Theses, https://doi.org/10.1007/978-3-030-48336-4

(continued)

Name	Maximal depth (diameter of conduit in m)	Length of conduit in m	Inlet offset (position of inlet of channel above the invert in m)	Outlet offset (position of outlet of channel above the invert in m)
C3 (W1–W2)	3.25	1500	0	0
C4 (W1–W2)	4	1500	10	0
C5 (W2–W3)	4.25	2000	0	0
C6 (W2–W3)	4	2000	23	0
C7 (W3–outflow)	3.8	2000	0	0

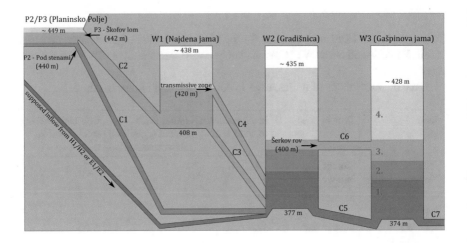

Appendix E

Conceptual model and settings used for the hydraulic model of the region with the cave H1 (Veliko brezno v Grudnovi dolini).

Some characteristics of nodes, used in conceptual and numerical model.

Name	Name	Type of node	Invert elevation of node in m	Initial depth (initial water level above invert in m)
Node H1 (Veliko brezno v Grudnovi dolini)	Junction with inflow	405	31	/
Node X1 (Hypothetical cave 1)	Storage unit	420	0	1000
Node X2 (Hypothetical cave 2)	Junction	395	0	/
Outflow	Outfall	350	0	/

Some characteristics of links, used in conceptual and numerical model.

Name	Maximal depth (diameter of conduit in m)	Length of conduit in m	Inlet offset (position of inlet of channel above the invert in m)	Outlet offset (position of outlet of channel above the invert in m)
C1 (H1–X1)	6	500	31	0
C2 (X1–X2)	3.8	2000	0	0
C3 (X1–X2)	1.75	2000	70	0
C4 (X2–outflow)	3.7	2000	0	0

© The Editor(s) (if applicable) and The Author(s), under exclusive license
to Springer Nature Switzerland AG 2020
M. Blatnik, *Groundwater Distribution in the Recharge Area
of Ljubljanica Springs*, Springer Theses, https://doi.org/10.1007/978-3-030-48336-4

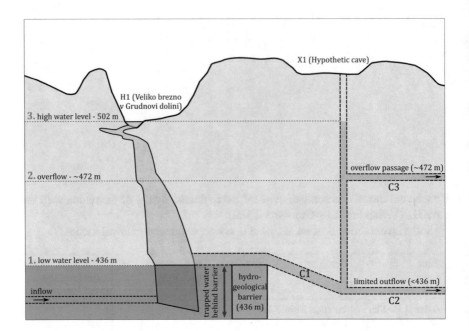

Appendix F

Conceptual model and settings used for the hydraulic model of the system between Planinsko Polje and caves H1 (Veliko brezno v Grudnovi dolini), H2 (Andrejevo brezno 1), and W3 (Gradišnica).

Some characteristics of nodes, used in conceptual and numerical model.

Name	Type of node	Invert elevation of node in m	Initial depth (initial water level above invert in m)	Constant (area of storage unit in m^2)
Node H1 (Veliko brezno v Grudnovi dolini)	Junction with inflow	405	31	/
Node X1 (Hypothetical cave 1)	Storage unit	420	0	1000
Node X2 (Hypothetical cave 2)	Junction	395	0	/
Node P1 (Planinsko Polje)	Storage unit with inflow	440	0	1,500,000
Node H2 (Andrejevo brezno 1)	Junction with inflow	435	0	/
Node W2 (Gradišnica)	Outfall	370	0	/

Some characteristics of links, used in conceptual and numerical model.

Name	Maximal depth (diameter of conduit in m)	Length of conduit in m	Inlet offset (position of inlet of channel above the invert in m)	Outlet offset (position of outlet of channel above the invert in m)
C1 (H1–X1)	6	500	31	0
C2 (X1–X2)	5.6	1000	0	0
C3 (P2–H2)	2.1	500	0	0
C4 (H2–X2)	2.1	100	0	0
C5 (X2–W2)	3.25	2000	0	0

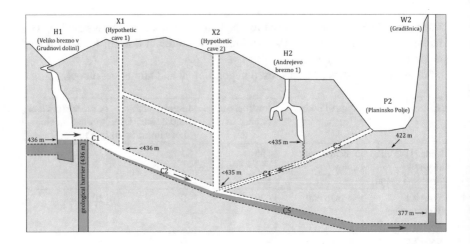

Appendix G

Flow rate of the inflow, used in models, presented in Appendices A–F.

Date	Inflow	Date	Inflow	Date	Inflow
05/13/2016 01:00	1.28	05/15/2016 05:00	42.76	05/17/2016 09:00	43.83
05/13/2016 02:00	2.26	05/15/2016 06:00	45.22	05/17/2016 10:00	43.47
05/13/2016 03:00	2.93	05/15/2016 07:00	47.26	05/17/2016 11:00	42.76
05/13/2016 04:00	3.60	05/15/2016 08:00	49.86	05/17/2016 12:00	42.40
05/13/2016 05:00	4.28	05/15/2016 09:00	51.72	05/17/2016 13:00	42.04
05/13/2016 06:00	4.28	05/15/2016 10:00	52.90	05/17/2016 14:00	41.31
05/13/2016 07:00	4.63	05/15/2016 11:00	53.76	05/17/2016 15:00	40.95
05/13/2016 08:00	4.97	05/15/2016 12:00	54.31	05/17/2016 16:00	40.58
05/13/2016 09:00	4.97	05/15/2016 13:00	54.86	05/17/2016 17:00	40.21
05/13/2016 10:00	5.32	05/15/2016 14:00	54.86	05/17/2016 18:00	39.84
05/13/2016 11:00	6.73	05/15/2016 15:00	55.13	05/17/2016 19:00	39.10
05/13/2016 12:00	8.53	05/15/2016 16:00	55.13	05/17/2016 20:00	38.72
05/13/2016 13:00	10.00	05/15/2016 17:00	55.13	05/17/2016 21:00	38.35
05/13/2016 14:00	11.86	05/15/2016 18:00	55.13	05/17/2016 22:00	37.97
05/13/2016 15:00	14.52	05/15/2016 19:00	55.13	05/17/2016 23:00	37.60
05/13/2016 16:00	17.23	05/15/2016 20:00	55.13	05/18/2016 01:00	36.46
05/13/2016 17:00	19.58	05/15/2016 21:00	55.13	05/18/2016 02:00	36.08
05/13/2016 18:00	21.55	05/15/2016 22:00	55.13	05/18/2016 03:00	35.69

(continued)

M. Blatnik, *Groundwater Distribution in the Recharge Area
of Ljubljanica Springs*, Springer Theses, https://doi.org/10.1007/978-3-030-48336-4

(continued)

Date	Inflow	Date	Inflow	Date	Inflow
05/13/2016 19:00	23.53	05/15/2016 23:00	55.13	05/18/2016 04:00	35.31
05/13/2016 20:00	25.51	05/16/2016 01:00	55.13	05/18/2016 05:00	34.93
05/13/2016 21:00	27.09	05/16/2016 02:00	55.13	05/18/2016 06:00	34.54
05/13/2016 22:00	28.28	05/16/2016 03:00	55.13	05/18/2016 07:00	34.15
05/13/2016 23:00	29.46	05/16/2016 04:00	54.86	05/18/2016 08:00	33.77
05/14/2016 01:00	30.64	05/16/2016 05:00	54.86	05/18/2016 09:00	33.38
05/14/2016 02:00	31.04	05/16/2016 06:00	54.59	05/18/2016 10:00	32.99
05/14/2016 03:00	31.82	05/16/2016 07:00	54.59	05/18/2016 11:00	32.60
05/14/2016 04:00	31.82	05/16/2016 08:00	54.31	05/18/2016 12:00	32.21
05/14/2016 05:00	32.60	05/16/2016 09:00	54.04	05/18/2016 13:00	31.82
05/14/2016 06:00	33.38	05/16/2016 10:00	53.76	05/18/2016 14:00	31.43
05/14/2016 07:00	34.15	05/16/2016 11:00	53.19	05/18/2016 15:00	31.04
05/14/2016 08:00	34.93	05/16/2016 12:00	52.90	05/18/2016 16:00	31.04
05/14/2016 09:00	35.69	05/16/2016 13:00	52.61	05/18/2016 17:00	30.25
05/14/2016 10:00	36.08	05/16/2016 14:00	52.31	05/18/2016 18:00	30.25
05/14/2016 11:00	36.46	05/16/2016 15:00	51.72	05/18/2016 19:00	29.86
05/14/2016 12:00	36.46	05/16/2016 16:00	51.41	05/18/2016 20:00	29.46
05/14/2016 13:00	36.46	05/16/2016 17:00	51.11	05/18/2016 21:00	29.46
05/14/2016 14:00	36.46	05/16/2016 18:00	50.80	05/18/2016 22:00	29.07
05/14/2016 15:00	35.69	05/16/2016 19:00	50.49	05/18/2016 23:00	28.67
05/14/2016 16:00	35.31	05/16/2016 20:00	49.86	05/19/2016 01:00	28.28
05/14/2016 17:00	34.93	05/16/2016 21:00	49.55	05/19/2016 02:00	27.88
05/14/2016 18:00	34.54	05/16/2016 22:00	49.23	05/19/2016 03:00	27.88
05/14/2016 19:00	33.38	05/16/2016 23:00	48.58	05/19/2016 04:00	27.49
05/14/2016 20:00	32.99	05/17/2016 01:00	47.93	05/19/2016 05:00	27.09
05/14/2016 21:00	32.21	05/17/2016 02:00	47.26	05/19/2016 06:00	27.09
05/14/2016 22:00	32.60	05/17/2016 03:00	46.93	05/19/2016 07:00	26.70
05/14/2016 23:00	33.38	05/17/2016 04:00	46.25	05/19/2016 08:00	26.70
05/15/2016 01:00	36.08	05/17/2016 05:00	45.91	05/19/2016 09:00	26.30
05/15/2016 02:00	37.60	05/17/2016 06:00	45.57	05/19/2016 10:00	26.30
05/15/2016 03:00	39.10	05/17/2016 07:00	44.88	05/19/2016 11:00	25.90
05/15/2016 04:00	40.58	05/17/2016 08:00	44.53	05/19/2016 12:00	25.90

Date	Inflow	Date	Inflow	Date	Inflow
05/19/2016 13:00	25.51	05/21/2016 17:00	17.23	05/23/2016 21:00	6.02
05/19/2016 14:00	25.51	05/21/2016 18:00	16.84	05/23/2016 22:00	6.02
05/19/2016 15:00	25.51	05/21/2016 19:00	16.84	05/23/2016 23:00	5.67
05/19/2016 16:00	25.11	05/21/2016 20:00	16.45	05/24/2016 01:00	5.32
05/19/2016 17:00	25.11	05/21/2016 21:00	16.45	05/24/2016 02:00	4.97
05/19/2016 18:00	25.11	05/21/2016 22:00	16.07	05/24/2016 03:00	4.97
05/19/2016 19:00	24.72	05/21/2016 23:00	16.07	05/24/2016 04:00	4.97
05/19/2016 20:00	24.72	05/22/2016 01:00	15.29	05/24/2016 05:00	4.63
05/19/2016 21:00	24.72	05/22/2016 02:00	15.29	05/24/2016 06:00	4.63
05/19/2016 22:00	24.32	05/22/2016 03:00	14.91	05/24/2016 07:00	4.63
05/19/2016 23:00	24.32	05/22/2016 04:00	14.91	05/24/2016 08:00	4.28
05/20/2016 01:00	23.92	05/22/2016 05:00	14.52	05/24/2016 09:00	4.28
05/20/2016 02:00	23.92	05/22/2016 06:00	14.14	05/24/2016 10:00	4.28
05/20/2016 03:00	23.92	05/22/2016 07:00	14.14	05/24/2016 11:00	3.94
05/20/2016 04:00	23.53	05/22/2016 08:00	13.76	05/24/2016 12:00	3.60
05/20/2016 05:00	23.53	05/22/2016 09:00	13.76	05/24/2016 13:00	3.60
05/20/2016 06:00	23.53	05/22/2016 10:00	13.38	05/24/2016 14:00	3.60
05/20/2016 07:00	23.53	05/22/2016 11:00	13.00	05/24/2016 15:00	3.26
05/20/2016 08:00	23.53	05/22/2016 12:00	13.00	05/24/2016 16:00	3.26
05/20/2016 09:00	23.13	05/22/2016 13:00	12.62	05/24/2016 17:00	2.93
05/20/2016 10:00	23.13	05/22/2016 14:00	12.62	05/24/2016 18:00	2.93
05/20/2016 11:00	23.13	05/22/2016 15:00	12.24	05/24/2016 19:00	2.93
05/20/2016 12:00	22.74	05/22/2016 16:00	12.24	05/24/2016 20:00	2.59
05/20/2016 13:00	22.74	05/22/2016 17:00	11.86	05/24/2016 21:00	2.59
05/20/2016 14:00	22.34	05/22/2016 18:00	11.49	05/24/2016 22:00	2.26
05/20/2016 15:00	22.34	05/22/2016 19:00	11.49	05/24/2016 23:00	1.93
05/20/2016 16:00	22.34	05/22/2016 20:00	11.11	05/25/2016 01:00	1.60
05/20/2016 17:00	21.94	05/22/2016 21:00	11.11	05/25/2016 02:00	1.60
05/20/2016 18:00	21.55	05/22/2016 22:00	10.74	05/25/2016 03:00	1.28
05/20/2016 19:00	21.55	05/22/2016 23:00	10.74	05/25/2016 04:00	1.28
05/20/2016 20:00	21.55	05/23/2016 01:00	10.00	05/25/2016 05:00	1.28
05/20/2016 21:00	21.15	05/23/2016 02:00	10.00	05/25/2016 06:00	0.95
05/20/2016 22:00	21.15	05/23/2016 03:00	9.63	05/25/2016 07:00	0.95
05/20/2016 23:00	20.76	05/23/2016 04:00	9.63	05/25/2016 08:00	0.64
05/21/2016 01:00	20.37	05/23/2016 05:00	9.26	05/25/2016 09:00	0.32
05/21/2016 02:00	20.37	05/23/2016 06:00	8.90	05/25/2016 10:00	0.32
05/21/2016 03:00	19.97	05/23/2016 07:00	8.90	05/25/2016 11:00	0.00
05/21/2016 04:00	19.97	05/23/2016 08:00	8.53	05/25/2016 12:00	0.00

(continued)

(continued)

Date	Inflow	Date	Inflow	Date	Inflow
05/21/2016 05:00	19.58	05/23/2016 09:00	8.53		
05/21/2016 06:00	19.58	05/23/2016 10:00	8.17		
05/21/2016 07:00	19.19	05/23/2016 11:00	8.17		
05/21/2016 08:00	19.19	05/23/2016 12:00	7.81		
05/21/2016 09:00	18.79	05/23/2016 13:00	7.81		
05/21/2016 10:00	18.40	05/23/2016 14:00	7.45		
05/21/2016 11:00	18.40	05/23/2016 15:00	7.09		
05/21/2016 12:00	18.01	05/23/2016 16:00	7.09		
05/21/2016 13:00	18.01	05/23/2016 17:00	6.73		
05/21/2016 14:00	17.62	05/23/2016 18:00	6.73		
05/21/2016 15:00	17.62	05/23/2016 19:00	6.38		
05/21/2016 16:00	17.23	05/23/2016 20:00	6.38		

Printed in the United States
by Baker & Taylor Publisher Services